GIANT OIL FIELDS AND WORLD OIL RESOURCES

PREPARED FOR THE CENTRAL INTELLIGENCE AGENCY

RICHARD NEHRING

R-2284-CIA
JUNE 1978

Rand
SANTA MONICA, CA. 90406

PREFACE

This report presents the results of Rand research to date on the size and distribution of the known and ultimately recoverable conventional crude oil resources of the world. The research was sponsored by the Office of Economic Research of the Central Intelligence Agency. The research project was initiated for two reasons: (1) to provide a detailed, publicly available description of the known recoverable crude oil resources of the world and (2) to provide an explicitly reasoned estimated range of ultimately recoverable conventional world crude oil resources. Both the description and the estimation are based primarily on a detailed study of the giant oil fields of the world. Estimates of ultimate world oil resources have considerable policy implications. Except for some brief comments in the concluding section, this report does not consider these implications. Rather, its purpose is only to provide a detailed description and estimation of known and ultimately recoverable conventional oil resources. Some of the policy implications are being examined in other Rand research, particularly the study of the future world oil market described below.

The research was undertaken as part of The Rand Corporation's International Security Policy Program under the direction of R. H. Solomon. It was conducted in conjunction with two other research projects in Rand's Energy Policy Program, directed by F. S. Hoffman. The first of these, sponsored by the Department of Energy, is an investigation of the future world oil market, using the data developed in this report. The second, sponsored by the U.S. Geological Survey, is an investigation of the implications of the discovery patterns of significant oil and gas fields in the United States for resource estimation.

This report should be of primary interest to those concerned with the question of ultimate oil resources and their significance for energy policy. It is also offered as a contribution to the literature on resource assessment and as a basic reference work on world oil.

SUMMARY

Expectations about future oil availability and ultimate conventional world oil resources play a major role in shaping energy policy proposals. However, considerable disagreement exists both about oil availability and ultimately recoverable conventional world oil resources, hampering efforts to reach agreement on energy policies. The absence of consensus is the result both of several economic, technical, and geologic uncertainties that cannot be currently resolved and of a failure to exploit existing public information about world oil resources.

This report is an initial attempt to remedy the latter problem, seeking to narrow the debate to the inescapable range of uncertainty. Its purpose is to assess the conventional known and prospective recoverable oil resources of the world primarily by describing and analyzing the giant oil fields of the world. Because these fields, usually defined as those with an ultimate recovery of 500 million barrels or more, contain more than 75% of the known recoverable oil resources of the world, a comprehensive examination of them provides an efficient means of assessing world oil resources.

This report lists the giant fields of the world and describes the means used to identify them and determine their size. It discusses several aspects of the worldwide distribution of known recoverable oil resources, focusing on the role of giant fields in each type of distribution. It describes the pattern of discoveries of giant fields since petroleum exploration began. The report concludes by estimating the range of ultimately recoverable conventional world oil resources and by discussing the considerations deemed by the author to be most significant for estimation. The data base used in the descriptions and analyses is provided in its entirety in Appendix A, together with explanatory notes on the quality and characteristics of the data and a list of the sources used in constructing the data base.

Section II describes the definitional and methodological considerations employed in this study to construct a list of giant oil fields and provides this list, together with a comparison with previous lists.

A *field* is defined as a producing area containing in the subsurface (1) a single pool uninterrupted by permeability barriers, (2) multiple pools trapped by a common geologic feature, or (3) laterally distinct multiple pools within a common formation and trapped by the same type of geologic separation where the lateral separation does not exceed one-half mile. *Giant oil fields* are all fields containing at least 500 million barrels of known recoverable crude oil. *Super-giant oil fields* are all fields containing at least 5 billion barrels of known recoverable crude oil. *Combination giant oil fields* are fields containing at least 250 million barrels of recoverable petroleum liquids and at least 500 million barrels of recoverable hydrocarbons in liquids or liquid-equivalents. Giant and combination giant oil fields constitute the *known giant oil fields*. *Potential giant oil fields* are fields for which some information exists to indicate that with development or additional recovery they may become known giant oil fields. Potential giants are divided into two categories: *probable giants* and *possible giants*, the distinction depending on the degree of certainty as to their potential for becoming giants.

To develop the list of known and potential oil fields, information was collected about all major recent discoveries and all fields in the world known to have an ultimate recovery of 200 million barrels or more. Information was sought for each field about the year of discovery, 1975 production, cumulative production to the end of 1975, and proved and probable reserves at the end of 1975. In order to test published estimates of cumulative production and reserves, information about the reservoir capacity, production and incremental recovery, and the economics of each field were sought as well. The sum of estimates of cumulative production, reserves, and total recovery for the giant fields in each country were checked with the country totals for these items as a second test of consistency.

At the end of 1975, there were 272 known giant oil fields in the world, 243 of which were full giant oil fields and 29 of which were combination giant oil fields. There were another 36 probable and 117 possible giant and combination giant oil fields. The information used in constructing this list was judged to be adequate for most descriptive and analytic purposes; however, the estimates of total recovery for most fields are somewhat uncertain, many being likely to increase with additional recovery resulting from higher returns to producers. This list is more comprehensive than the other two lists published in the past decade. However, it excludes or downgrades in status several fields included in those lists either for definitional reasons or because the available information indicated that the fields were not giants.

Section III describes several aspects of the composition of known conventional world oil resources. The distribution of known recoverable resources by size of field, by country, and by province is outlined and briefly analyzed. In each case, the significance of giant fields is emphasized. Including both proved and probable reserves, the known recoverable crude oil resources of the world at the end of 1975 were 1011.5 billion barrels, 335.1 of which have already been produced. This estimate is compared with four other recent, widely used estimates. Excluding conceptual differences (primarily the extent to which each estimate includes probable reserves), the major differences between this report and the others are the estimates for the Soviet Union and Indonesia, this report arguing for lower reserves in each case.

Known recoverable crude oil resources are heavily concentrated in giant fields. An estimated 819.2 of the 1011.5 billion barrels are contained in known and potential giant fields. Over half of the total is in the 33 super-giant fields. Roughly 10% of the total is found in the estimated 20,000 plus fields smaller than 100 million barrels each. The distribution by size of field indicates unmistakably that the growth of known recoverable world oil resources to their present level has been dependent on the discovery and development of giant and particularly super-giant oil fields.

World oil resources are also concentrated within a few countries. The seven largest--Saudi Arabia, the United States, the Soviet Union, Iran, Kuwait, Venezuela, and Iraq--have nearly three-fourths of the known resource. The 17 countries with 10 billion barrels or more each have 93% of the known resource. The distribution of oil resources by country is principally dependent on the distribution of giant oil fields. The four largest countries have half of the known giant oil fields. All of the countries with 10 billion barrels or more have at least five known or potential giant oil fields. All but three of the countries with no known or potential giant oil fields have a total recovery of less than 500 million barrels.

World oil resources are also heavily concentrated within a few producing provinces, that is, sets of oil and gas fields that are geographically contiguous and that occur in a similar or related geological environment. The 22 major provinces discovered to date (those that have or are likely to have a total recovery of 10 billion barrels or more) contain 86% of the known recoverable resources. This predominance stems from the fact that they contain all of the super-giant fields and 95% of the recoverable oil in known giant fields. Of the 275 provinces that have been explored to some degree, production or significant discoveries have occurred in 125, 68 of which have known or potential giant oil fields. The Arabian-Iranian Basin (in the Middle East) is clearly unique as a province. It contains more than half of the known recoverable oil of the world, a predominance resulting primarily from the fact that it has 76% of the super-giant fields.

Additions to world oil resources in the past have primarily been a function of the rate of discovery of giant oil fields. Section IV describes the rate of discovery of giant oil fields to 1975, both worldwide and in selected regions. Before 1925, the number of giant fields discovered in each five-year period was at a low level, averaging about one per year. With the development of exploratory technology and a growing sophistication in geologic thinking, the *number* of discoveries has accelerated since then (although this trend was interrupted by the Great Depression and by World War II). The number of known and probable giant fields discovered from 1961 to 1975 averaged 9 to 10 per year. The *amount* discovered in giant fields reached a high plateau between the late 1930s and the early 1960s (except for World War II). Since the early 1960s it has dropped precipitously. The decrease in the amount discovered (roughly 40% to 55% after adjusting for likely reserve growth) is primarily the result of the sharp drop in the number of super-giant fields discovered in each five-year period. In the early 1960s, eight recognized and five potential super-giants were discovered. In the early 1970s, at most two potential ones were discovered.

The effect of the decline in the discovery of super-giants is most apparent in the Middle East, where the amount discovered in giant fields has dropped sharply since the early 1960s. Outside the Middle East, the amount discovered in giant fields in each of the past four five-year periods has been at least 40 to 50 billion barrels. This high plateau has been the result of the continued discovery and development of new major provinces. Because it now generally takes only five to ten years to discover the major fields in a province, locating new major provinces appears to be essential for a continued level of major discoveries. Since discoveries of giant fields in North America reached their peak in the late 1920s, further discoveries (particularly since World War II) have nearly all resulted from the exploration and development of new provinces. The relatively short length of the period of major exploratory success is also illustrated by the pattern of discovery in Africa since 1950. Both the amount discovered and the number of giants discovered quickly reached a peak after exploration began. Since the peak of the late 1950s and early 1960s, major exploratory successes in Africa have declined sharply.

Section V provides an explicit evaluation of ultimately recoverable conventional oil resources, based primarily on an analysis of the potential contributions of giant oil fields. Both future discoveries and additional recovery from known fields are considered. The reasoning behind the judgmental estimates is provided in some detail so that the reader can understand how the estimates were derived and can check them against other possibilities.

The potential for further discoveries is evaluated through a consideration of the possibilities for the discovery of new major provinces, super-giant fields, other giant fields, and non-giant fields. If more new major provinces are to be discovered, they are most likely to be relatively unexplored intracontinental-composite, extracontinental, and graben or rift provinces with a large surface area. Considering the declining rate of discovery of new major provinces, the historic probabilities that a province of a specific area and type will be a major province, and the number and characteristics of the remaining possibilities, the author estimates that only two to four more major provinces will be discovered and developed.

It appears unlikely that many super-giant oil fields will be discovered in the future. The Arabian-Iranian and Reforma provinces are the only known producing provinces with a reasonable probability of further super-giant discoveries. Only one to two more are likely to be found in new provinces. A total of four to ten super-giants containing 30 to 100 billion barrels is estimated as remaining to be discovered.

The discovery rate for other giant fields will probably decline as the primary discovery process in the known major provinces is completed and few new major provinces are found. About two-thirds of the known major provinces and some of the lesser producing provinces still have some potential for future giant field discoveries, the best possibilities being in the Arabian-Iranian and Reforma provinces. A total of 125 to 175 other giant fields containing 100 to 175 billion barrels is estimated as remaining to be discovered.

With the possible exceptions of substantial numbers of future discoveries in subtle traps and in the Arabian-Iranian province, future trends in the discovery of large non-giant fields are postulated to be roughly similar to those of the other giant fields. A total of 400 to 800 large non-giant fields containing 60 to 160 billion barrels is estimated as remaining to be discovered. Except for the Middle East, the proportion discovered in the smaller non-giant fields in the producing provinces of the world is projected to become more similar to that in the United States and Canada. A total of 73 to 120 billion barrels in these fields is estimated as remaining to be discovered. The total amount remaining to be discovered worldwide is estimated to be 263 to 555 billion barrels.

In the future, additional recovery from known fields is likely to be more important for additions to total recovery worldwide than new discoveries. Assuming that world oil prices will increase to at least $30 per barrel (constant dollars), ultimate recovery in the United States is likely to be between 40% and 50% of original oil-in-place. Because of the relatively unfavorable distribution of reservoir characteristics in the United States and the early history of U.S. oil production under wasteful institutional arrangements, recovery rates elsewhere in the world are likely to be even higher. The potential for additional recovery in the Middle East is particularly promising. The combined potential for reserve growth in known fields from extensions, new-pool discoveries, full development, and additional recovery is estimated to range from 420 to 730 billion barrels.

The ultimately recoverable conventional crude oil resources of the world are estimated to be 1700 to 2300 billion barrels. The probability that the ultimate amount will be less than 1700 billion barrels is quite low. The ultimate amount could be more than 2300 billion barrels either if large non-analogous discoveries are made or if additional recovery efforts prove to be highly successful. The former possibility should be monitored as exploration continues. The latter is likely to become apparent only in the next century. Neither the distribution by field size nor by region of ultimate conventional world oil resources is estimated to be significantly different from the respective distributions of known conventional oil resources. The rate at which the ultimate resource will become available depends primarily on the development of technology for offshore Arctic and deep-water exploration and production, the production policies of the OPEC countries, and the existence of the necessary economic incentives to producers and refiners.

ACKNOWLEDGMENTS

I would like to thank the staff of the Office of Economic Research for their helpful support of this project. F. S. Hoffman, M. Kennedy, and C. Phelps of Rand reviewed the preliminary drafts of this report. Their suggestions helped considerably in sharpening the arguments and improving the clarity of the manuscript. I would also like to thank those individuals in industry and government who generously provided data and reviewed the initial draft, E. T. Gernert for editing the manuscript, and M. Schubert for her work in producing the appendix.

CONTENTS

FIGURES

TABLES

I. INTRODUCTION

Expectations about future oil availability and ultimate conventional* world oil resources play a major role in energy policy proposals worldwide. This role is readily understandable for several reasons. Crude oil is the largest primary source of energy worldwide, providing between 40% and 50% of world energy use. Its use has grown exponentially for more than a century, increasing at an average annual rate of 7% from the early 1860s to the early 1970s. (At this rate of growth, both annual and cumulative use doubles every decade.) Crude oil is a nonrenewable resource, making the consideration of its depletion and substitution an essential element in any energy policy.

Because of the central role of oil in world energy use, different expectations about future oil availability underlie many disagreements about energy policy proposals. No consensus exists, either about future oil availability (i.e., the path of annual production) or about ultimate recoverable conventional world oil resources. Since 1970, published estimates of the latter have varied between 1.2 and 4.0 trillion barrels. (As the difference between the two extremes is equal to more than 100 years of world oil consumption at current rates, the implications of this disagreement for policy are obvious.) On the more immediate questions of near-term availability, charges that reserve estimates are understated, that production is being withheld to maintain or boost prices, or that exploration has been deliberately inefficient to give a misleading impression of resource scarcity have been given much media attention.

---

*Conventional oil resources are only one of two subsets of nature-made oil resources. The other, nonconventional oil resources, consists of oil (tar) sands and oil shale. There is no uniform dividing line between oil (tar) sands and conventional heavy oil. R. F. Meyer and C. R. Hocott ("Summary," in R. F. Meyer (ed.), *The Future Supply of Nature-Made Petroleum and Gas*, Pergamon Press, New York, 1977, pp. 16-17) have suggested that all reservoirs containing oil heavier than 7° to perhaps 10° API gravity should be classified as oil (tar) sands. The criterion of 8° to 10° API is used in this report, permitting the inclusion of heavy oil fields in which production by conventional methods has already occurred.

Some skepticism of current reserve estimates worldwide is appropriate. With few exceptions (such as the excellent annual reports of the Provinces of Alberta and Saskatchewan in Canada and the States of Montana and North Dakota in the United States), detailed comprehensive reservoir information by field is not available. The economic assumptions in reserve calculations are never made very clear by any source. Information on the range of technical possibilities is also not readily available.

Estimates of ultimate world resources are documented even less well. Only a few are disaggregated by country or region; only a few explicitly incorporate uncertainty; and only a few describe the reasoning behind the estimates in any detail. In general, what is presented are largely "naked" numbers, devoid of any convincing or even testable rationale, and one is presumably expected to believe them on the basis of the claimed expertise of the individual or organization who made the estimate.

Theoretically, some facts about world oil resources are amenable to precise, explicit determination. A comprehensive, reliable estimate of the known recoverable oil resources of the world could be made through a field-by-field determination of oil-in-place based on the uniform application of engineering calculations, a determination of primary and secondary recovery in each field assuming uniform economic conditions and current technology, and an analysis of enhanced recovery possibilities in each field and their costs. But even if the basic field data were publicly available to permit such an assessment (and they are not), the task would be one of staggering dimensions and cost, given that there are 20,000 to 30,000 oil fields in the world.

The data problem can be greatly simplified without a substantial loss of accuracy or information by focusing on the giant oil fields of the world.* These fields, only several hundred in number, contain

---

*Giant oil fields are usually defined as fields with an ultimate recovery of 500 million barrels or more. For the literature on giant fields, see G. M. Knebel and G. Rodriguez-Eraso, "Habitat of Some Oil," *American Association of Petroleum Geologists Bulletin*, Vol. 40, No. 4, April 1956, pp. 547-561; R. J. Burke and F. J. Gardner, "The World's Monster Oil Fields and How They Rank," *Oil and Gas Journal*, Vol. 67, No. 2, January 13, 1969; M. T. Halbouty (ed.), *Geology of Giant Petroleum Fields*, American Association of Petroleum Geologists Memoir 14,

more than three-fourths of the known recoverable oil resources of the world. Because of their prominence, both worldwide and within the countries in which they are found, in general more information is available about them than about other oil fields. A comprehensive examination of giant oil fields thus provides the basis for an efficient disaggregated assessment of known recoverable world oil resources. Because they are relatively small in number, yet contain the bulk of the world's oil resource, the costs of data acquisition and verification about them are reasonably low, and the problems of aggregating and analyzing the data are manageable.

The purpose of this report is to assess the conventional known and prospective recoverable oil resources of the world, primarily by

---

Tulsa, Okla., 1970, especially M. T. Halbouty et al., "World's Giant Oil and Gas Fields, Geologic Factors Affecting Their Formation and Basin Classification," pp. 502-555; H. D. Klemme, "The Giants and the Supergiants," *Oil and Gas Journal*, Vol. 69, Nos. 9, 10, and 11, March 1, 8, and 15, 1971; H. D. Klemme, "Trends in Basin Development: Possible Economic Implications," *World Petroleum*, October 1971; H. D. Klemme, "Structure-Related Traps Expected To Dominate World Reserve Statistics," *Oil and Gas Journal*, Vol. 71, No. 53, December 31, 1973, and Vol. 72, No. 1, January 7, 1974; H. D. Klemme, "Giant Oil Fields Related to Their Geologic Setting: A Possible Guide to Exploration," *Bulletin of Canadian Petroleum Geology*, Vol. 23, No. 1, March 1975, pp. 30-66; J. D. Moody, "Distribution and Geological Characteristics of Giant Oil Fields," in A. G. Fischer and S. Judson (eds.), *Petroleum and Global Tectonics*, Princeton University Press, Princeton, N.J., 1975, pp. 307-320; J. D. Moody and R. W. Esser, "An Estimate of the World's Recoverable Crude Oil Resource," *Proceedings, Ninth World Petroleum Congress*, Vol. 3 (Exploration and Transportation), Applied Science Publishers, London, 1975, pp. 11-20; D. A. Holmgren, J. D. Moody, and H. H. Emmerich, "The Structural Settings for Giant Oil and Gas Fields," *Proceedings, Ninth World Petroleum Congress*, Vol. 2 (Exploration and Transportation), Applied Science Publishers, London, 1975, pp. 45-54; A. A. Meyerhoff, "Economic Impact and Geopolitical Implications of Giant Petroleum Fields," *American Scientist*, Vol. 64, No. 5, September-October 1976, pp. 536-541; H. D. Klemme, "Giant Fields Contain Less Than 1% of World's Fields But 75% of Reserves," *Oil and Gas Journal*, Vol. 75, No. 10, March 7, 1977; H. D. Klemme, "One-Fifth of Reserves Lie Offshore," *Oil and Gas Journal, Petroleum/2000*, Vol. 75, No. 35, August 1977; H. D. Klemme, "World Oil and Gas Reserves from Analysis of Giant Fields and Petroleum Basins (Provinces)," in R. F. Meyer (ed.), *The Future Supply of Nature-Made Petroleum and Gas*, pp. 217-260; and G. J. Demaison, "Tar Sands and Supergiant Oil Fields," *American Association of Petroleum Geologists Bulletin*, Vol. 61, No. 11, November 1977, pp. 1950-1961. Lists of the large oil fields of the world, mostly giants, are also published annually in the *International Petroleum Encyclopedia*, Petroleum Publishing Company, Tulsa, Okla.

means of a description and analysis of the giant oil fields of the world. The report lists these giant fields and describes the means used to identify them and determine their size. It describes several aspects of the worldwide distribution of known recoverable oil resources, focusing on the role of giant fields in each type of distribution. It describes the pattern of discoveries of giant fields since petroleum exploration began. It concludes by estimating the range of ultimately recoverable conventional world oil resources. The reasoning behind the judgmental estimates is provided in some detail so that the reader can check them against the supporting data and compare them with other estimates.

The data base used in the descriptions and analyses is provided in its entirety in Appendix A, together with explanatory notes on the quality and characteristics of the data and a list of the sources used in constructing the data base. Although this appendix was derived from a critical and reasonably thorough use of the available public sources of information, it carries no claim of being a wholly accurate and uniform description of the known recoverable oil resources of the world. As the explanatory notes indicate, such a description is not possible now even if one had ready access to all existing information, publicly available or not, about the discovered oil fields of the world. What the appendix does provide is a complete, systematically organized set of the disaggregated data used to formulate the descriptions and conclusions of this report. The data are published in their entirety to enable the reader (1) to determine whether the descriptions and conclusions are adequately supported by the data, (2) to formulate other descriptions and conclusions that may be more in agreement with the data, and (3) to compare the data used here with data from other, possibly conflicting sources, checking the documentation and determining whether the descriptions and conclusions of this report are robust enough to survive amendments in the basic data.[*]

---

[*] For a statement on the need for explicitness in resource estimation, see J. D. Haun, "Methods of Estimating the Volume of Undiscovered Oil and Gas Resources--AAPG Research Conference," in J. D. Haun (ed.), *Methods of Estimating the Volume of Undiscovered Oil and Gas Resources*, American Association of Petroleum Geologists Studies in Geology, No. 1, Tulsa, Okla., 1975, p. 7.

II. THE GIANT OIL FIELDS OF THE WORLD

METHODOLOGY

Analyzing known and potential world oil resources by means of an analysis of giant fields requires a comprehensive list of giant fields and accurate information about them. Compiling a list that meets these criteria is not an easy task. This section describes the definitional and methodological considerations employed in this study to construct a complete and consistent list of giant oil fields and to determine accurate information about their oil resources. The discussion of methodological issues is followed by the list of giant fields developed for this report, and this list is compared briefly with previous lists.

Field Definition

The term *field* is used to refer to a single accumulation (in which case it is synonymous with a *pool* or *reservoir*) or a set of closely related accumulations of petroleum.* Although there is agreement on this general definition, the term is not used uniformly among or even within countries, partly because of the historical and institutional circumstances of discovery and development and partly because of disagreement as to what constitutes a *closely* related set of accumulations.

As used in this report, a *field* is a producing area containing in the subsurface (1) a single pool uninterrupted by permeability barriers, (2) multiple pools trapped by a common geologic feature, or (3) laterally distinct multiple pools within a common formation and trapped by the same type of geologic feature where the lateral separation between pools does not exceed one-half mile. This definition includes sets of overlapping pools of different geologic ages within a single geologic feature such as an anticline, sets of tangential pools where the pools are separated by impermeable faults, single pools that have several

---

*See A. I. Levorsen, *Geology of Petroleum*, 2d ed., W. H. Freeman and Company, San Francisco, 1967, p. 30; and G. D. Hobson and E. N. Tiratsoo, *Introduction to Petroleum Geology*, Scientific Press, Beaconsfield, England, 1975, p. 6.

field designations for administrative purposes, and sets of adjacent pools around and above the same salt dome. Smaller pools discovered in the process of developing a large stratigraphic pool are also considered to be in the same field as the original pool. The definition excludes sets of laterally distinct adjacent pools (1) that are combined as a single field for administrative purposes, (2) that share a common regional geological feature but are more than one-half mile apart, or (3) that are separated by a distinctive geologic feature (e.g., the channel in the Horseshoe Atoll between *Cogdell* and *Scurry* (*Kelly-Snyder-Diamond M*) fields in the Permian Basin in west Texas).[*]

Field Classification

Giant oil fields are defined in this report as all fields from which at least 500 million barrels of crude oil have been or are expected to be recovered. This usage thus continues the definition of a giant oil field that has become the accepted usage during the past decade.[†] Other definitions have been used. Knebel and Rodriguez-Eraso, in their pathbreaking analysis of large oil fields worldwide,

---

[*]This definition is a modified version of the one used by Halbouty et al., "World's Giant Oil and Gas Fields," p. 504. Their definition is modified to provide for more consistent usage consonant with generally accepted field nomenclature. If the criteria they used in practice to designate several sets of pools as fields were applied consistently on a worldwide basis, the resulting list of giant fields would incorporate a methodological bias in favor of designating sets of large accumulations as giant fields.

[†]This definition appears to have been used first by J. D. Moody, J. W. Mooney, and J. Spivak, "Giant Oil Fields of North America," in Halbouty (ed.), *Geology of Giant Petroleum Fields*, pp. 8-17. It has been subsequently used, among others, by Halbouty et al., "World's Giant Oil and Gas Fields"; Moody and Esser, "An Estimate of the World's Recoverable Crude Oil Resource"; Holmgren, Moody, and Emmerich, "The Structural Settings for Giant Oil and Gas Fields"; Meyerhoff, "Economic Impact and Geopolitical Implications of Giant Petroleum Fields"; L. F. Ivanhoe, "Evaluating Prospective Basins," *Oil and Gas Journal*, Vol. 74, Nos. 49, 50, and 51, December 6, 13, and 20, 1976; and Klemme, "Giant Fields Contain Less Than 1% of World's Fields But 75% of Reserves." This continuity in usage is in part explained by the fact that most of the subsequent literature on giant fields has been written by the same people who were responsible for the initial analyses.

used 100 million barrels as a minimum level.* This level is still commonly used as the minimum level for giant oil fields in the United States.† One of the earliest listings of large oil fields worldwide used 1 billion barrels as a minimum level.‡ The minimum of 500 million barrels is used here because it facilitates consistent comparisons with the bulk of the literature on giant fields, because it captures the bulk of known world oil resources without creating major problems of data acquisition, and because it is exactly one order of magnitude larger than the minimum size level of the largest field size category used by the American Association of Petroleum Geologists (Class A: 50 million barrels or more).

Within the giant classification, there is an important subclassification, that of *super-giant oil fields*. In this report, super-giant oil fields are defined as all fields that contain at least 5 billion barrels of known recoverable crude oil. This level is chosen primarily because it is exactly one order of magnitude larger than the minimum size level for giant fields. Although this subclassification has been used in the literature, no definitional consensus has emerged. Both 4 billion and 10 billion barrels have been proposed as minimum levels for super-giant fields.** Because usage has not been uniform, because there have been no reasons given for the definitions that have been proposed, and because the level of 5 billion barrels is symmetrical with other commonly used systems of field size classification, a third definition is proposed and used in this report.

Giant oil fields are frequently analyzed in conjunction with giant gas fields (all fields containing at least 3 trillion cubic feet of

---

*"Habitat of Some Oil."

†For example, M. T. Halbouty, "Giant Oil and Gas Fields in the United States," in Halbouty (ed.), *Geology of Giant Petroleum Fields*, pp. 91-127; and in "Giant Fields Still Yield Most U.S. Oil," *Oil and Gas Journal*, Vol. 76, No. 5, January 30, 1978 (see also previous years).

‡Burke and Gardner, "The World's Monster Oil Fields and How They Rank."

**Klemme, "The Giants and the Supergiants," used the former. Meyerhoff, "Economic Impact and Geopolitical Implications of Giant Petroleum Fields," used the latter.

recoverable natural gas, using the common gas-oil equivalency standard of 6000 cubic feet of dry natural gas to 1 barrel of crude oil). Although giant gas fields are not considered in this report, large recoverable natural gas resources are of interest in one respect. There are several large fields worldwide that do not have sufficient recoverable crude oil or natural gas resources to qualify as giant oil or gas fields. However, the sum of recoverable hydrocarbons of the different types in these fields may be of giant dimensions. This report uses the term *combination giants* to refer to such fields.* A combination giant oil field is defined as a field containing (1) at least 250 million barrels of recoverable petroleum liquids (crude oil and natural gas liquids), that is, at least half the minimum size of a giant oil field; and (2) at least 500 million barrels of recoverable hydrocarbons in liquids or liquid equivalents (natural gas being converted at 6000 cubic feet per barrel). Fields that meet the latter criterion but not the former are considered to be combination giant gas fields and thus are excluded in this report.

Because of a lack of sufficient information, previous compilations of giant oil fields have failed to be comprehensive; that is, they did not include all the fields now known to be giants that were discovered by the time the compilations were made.† Overcoming this problem appears to be an impossible task. At any point in time, there will be fields that have been discovered but are not yet developed or even delineated. Fields that have begun to produce may still be subject to further extensive development (e.g., extensions and new-pool discoveries) and intensive development (e.g., infill drilling, fracturing, acidizing). This is particularly likely to occur in those fields where production

---

*Meyerhoff, "Economic Impact and Geopolitical Implications of Giant Petroleum Fields," has used the concept without the term. However, no list of combination giant fields has been published before this report.

†For example, Knebel and Rodriguez-Eraso, "Habitat of Some Oil," listed 51 giant oil fields as of the end of 1955 in the world excluding the Soviet Union, Eastern Europe, and China. The list compiled for this report counted 90 giant oil fields, 19 combination giant oil fields, and 38 potential giant oil fields discovered as of the end of 1955 in the same group of countries.

has been constrained by regulatory controls, lack of sufficient demand, or higher relative costs. Secondary and enhanced recovery operations may also be implemented or expanded in subsequent years. Because of increases in total recovery for any of these reasons, some fields will not be recognized as giants until several years after their discovery.

In order to incorporate the phenomenon of reserve growth, this report uses the concept of *potential giants*, that is, fields that are not currently recognized as giant or combination giant oil fields but that may eventually be recognized as such. Potential giant oil fields are contrasted with *known giant oil fields*, that is, fields for which sufficient information is currently available to identify them as giant or combination giant oil fields. Potential giant fields may be either (1) fields for which no estimate of total recovery is currently available but for which some information exists indicating that they could ultimately prove to be giants, or (2) fields for which estimates of total recovery are currently available that are less than the amount required to qualify them as giants or combination giants, but for which some information exists to suggest that possible further development or implementation and expansion of secondary or enhanced recovery operations could ultimately qualify them for giant status.

Potential giant fields will be divided into two categories: *probable giants* and *possible giants*. The difference between the two is simply one of greater certainty. Probable giants are those fields for which sufficient information exists to indicate a high probability that they will be recognized as giants, most likely by 1990. Possible giants are those fields for which some information exists to indicate that they have the potential to become giants, but the quantity and quality of this information are less conclusive than that available for probable giants. The dividing line between the two is clearly a matter of judgment and the availability of information. Additional information may exist, which could not be obtained during the research for this report, that would suggest shifting individual fields between these two categories, shifting them from the status of potential giants to known giants, or deleting them from the list of known and potential giant fields.

A final category deserves brief mention. Because the importance of giant fields to world oil resources has become recognized during the past 20 years, some companies and countries have been eager to identify new discoveries as giant oil fields. In some cases, they have been overeager, subsequent development indicating that the discovery was clearly not of giant proportions. Such fields are designated in this report as *erstwhile giants*.

Data Quality

In developing the list of giant fields, both major recent discoveries and all fields in the world known to have an ultimate recovery of 200 million barrels or more were surveyed. For each field, information was sought about its year of discovery (defined as the year in which the first well was completed in the field), its 1975 production, its cumulative production to the end of 1975, and its proved and probable reserves at the end of 1975 (the sum of the latter two indicating its estimated total recovery and hence its size). Information about current and cumulative production and reserves of natural gas and natural gas liquids in these fields was also sought in order to obtain a comprehensive list of combination giant fields.

Proved reserves of crude oil are generally defined to be the quantity of crude oil demonstrated with reasonable certainty to be recoverable in future years from known reservoirs under existing economic and operating conditions. Probable reserves are likely additions to proved reserves from known fields not yet producing, from extensions and new-pool discoveries in known fields, and from implementation or expansion of secondary and enhanced recovery operations in known fields under existing economic and operating conditions. Theoretically, this should include all the oil that could be recovered with the use of existing technology at current world oil prices. In practice, published estimates of proved and probable reserves generally include only those quantities of crude oil expected to be recoverable in future years from investments in production facilities that have already been made, that are being made, or that are planned to be made during the next several years. These estimates thus indicate only a "working inventory" of recoverable oil resources. They, together with cumulative production, do not necessarily indicate the amounts that will ultimately be produced from known fields. Quantities generally excluded from published estimates of proved and probable reserves are those (1) that could be recovered from investment that is possible under current economic conditions but is not planned, (2) that could be made at current world oil prices but neither at current prices less economic rents (taxes and royalties) nor at current controlled prices, and (3) that could be made at plausible higher future prices. Because of the highly uneven availability of information about recoverable oil resources in these latter categories, the reserve estimates for countries and fields in this report correspond to present practice in order to maintain consistency among countries.*

Because some of the publicly available information about cumulative production, reserves, and total recovery is of doubtful quality or is conflicting, because such information was not available about some discoveries in structures of giant potential, and because there are substantial differences in the relevant economic conditions under which fields have been developed (or have not been developed), data relating to reservoir capacity, production and incremental recovery, and economic factors were also collected to provide a means of testing the published estimates of cumulative production and estimated reserves. Within each of these general categories of information, several specific items were sought, including:

Reservoir Capacity: Oil-in-place, field area, extent to which the field is delineated, average net thickness of the producing formation(s), reservoir porosity and permeability, formation volume factor, and initial oil saturation.

---

*For a dated but still highly useful discussion of various concepts of oil reserves, see W. F. Lovejoy and P. T. Homan, *Methods of Estimating Reserves of Crude Oil, Natural Gas, and Natural Gas Liquids*, The Johns Hopkins Press, Baltimore, Md., 1965. Also of interest is M. A. Adelman's discussion of the inventory concept of reserves in *The World Petroleum Market*, The Johns Hopkins Press, Baltimore, Md., 1972, pp. 24-42.

Production and Incremental Recovery: Annual production during the past five or more years, average decline rates, current or planned production capacity, constraints on production (regulatory, market, disruptions), projected producing life, development practices of the operating company, existence of undeveloped potentially productive zones, major technical production problems, estimated recovery factors, current or planned injection capacity, water cuts, problems in incremental recovery operations, enhanced recovery possibilities, and gas-oil ratios.

Economics: Investment costs per barrel of daily capacity, average production per well per day, test flows of discovery and extension wells, operating costs per daily barrel, total costs and revenues per barrel to operators, depth of wells, well spacing, water depth (for offshore fields), offshore operating conditions, accessibility of field, and oil quality data (gravity and viscosity, sulfur and metals content).

In no case was all this information available for any of the fields surveyed. For nearly all of the fields examined, several of these items of information were available to test the published estimates of cumulative production, reserves, and total recovery. Items of information that were considered crucial for the estimate of total recovery are generally included in the notes to the various tables in Appendix A. Very brief notes about specific fields generally indicate consistency between the supplemental information and published estimates of total recovery. For each country with giant fields, the summed estimates of cumulative production, reserves, and total recovery in these fields were checked with country totals of the same items as a second test of consistency. When no estimates of reserves or total recovery were available (generally in fields that have not yet been developed) or when the available estimates were judged to be highly unreliable, the other available information was used to make a conservative estimate of total recovery for the field if this information was sufficient to indicate that the field was a giant. When the information was insufficient to indicate that the field was a giant but sufficient to indicate that it might be one, the field was indicated as only a potential giant, often with no estimate of reserves or total recovery.

GIANT OIL FIELDS OF THE WORLD

Table 2.1 lists the known and probable giant oil fields of the world as of December 31, 1975, and provides information about their location by country, year of discovery, 1975 production, cumulative production, and reserves as of December 31, 1975, and estimated total recovery. If there is some uncertainty about the estimates of total recovery, that uncertainty is indicated as well. These fields, together with possible giant fields, are also listed by country in Appendix A. The tables of Appendix A also contain brief explanatory notes about the fields and the overall quality and reliability of the information about them, list the principal sources used in compiling the lists of known and potential giant fields, and indicate the range of recently published estimates of total recovery for most of these fields. The range of estimates of total recovery is provided only to indicate what other sources have said about particular fields since 1970. It does not indicate the current range of uncertainty. The full range of published estimates is included for comparative purposes, even if parts of that range are wholly implausible given currently available information.

At the end of 1975, there were 272 known giant oil fields in the world. Of these, 243 were full-fledged giant oil fields and the other 29 were combination giant oil fields. Another 36 fields were identified as probable giants or combination giants from the available information. There were also 117 possible giant oil fields, which are only listed in the tables of Appendix A. In 1976 and 1977, at least 12 other fields have been discovered that may prove to be giant fields.

Although the available information about reserves and total recovery of the known and potential giant fields is adequate to make useful summaries and analyses, the estimates shown are somewhat uncertain. Only 39 of the 272 known giant fields listed were judged not to have a reasonable probability of a change in estimates of total recovery of 10% or more during the next decade. The availability of information

Table 2.1

GIANT OIL FIELDS OF THE WORLD AS OF DECEMBER 31, 1975

(In millions of barrels)

Field	Country	Discovery Year	1975 Production	Cumulative Production	Reserves	Total Recovery
Giants						
1. Ghawar	Saudi Arabia	1948	1,665.6	12,542	70,458	83,000 (low?)
2. Burgan	Kuwait	1938	550.0^{E}	14,840^{E}	57,160	72,000 (?)
3. Bolivar Coastal*	Venezuela	1917	566.0^{E}	20,200^{E}	11,800	32,000 (low?)
4. Safaniya-Khafji**	Saudi Arabia/ Neutral Zone	1951	354.7^{E}	4,670	25,330	30,000 (low?)
5. Rumaila	Iraq	1953	355.0^{E}	2,523^{E}	17,477	20,000 (low?)
6. Ahwaz	Iran	1958	374.3	1,810	15,690	17,500 (?)
7. Kirkuk	Iraq	1927	350.0^{E}	6,933^{E}	9,067	16,000 (?)
8. Marun	Iran	1964	434.7	2,627	13,373	16,000 (?)
9. Gach Saran	Iran	1928	248.6	3,716	11,784	15,500 (?)
10. Agha Jari	Iran	1938	302.9	6,205	7,795	14,000 (?)
11. Samotlor	U.S.S.R.	1966	635.0	1,627	11,373	13,000 (?)
12. Abqaiq	Saudi Arabia	1940	281.3	5,058	7,442	12,500 (low?)
13. Romashkino	U.S.S.R.	1948	580.0^{E}	8,800^{E}	3,600	12,400 (?)
14. Berri*	Saudi Arabia	1964	153.8	885	11,115	12,000 (low?)
15. Zakum**	Abu Dhabi	1964	88.6	667	11,333	12,000 (?)
16. Manifa**	Saudi Arabia	1957	8.2	131	10,869	11,000 (low?)
17. Fereidoon-Marjan**	Iran/Saudi Arabia	1966	29.1	38	9,962	10,000 (?)
18. Prudhoe Bay	U.S.A.	1968	0.7	2	9,598	9,600 (low?)
19. Bu Hasa	Abu Dhabi	1962	170.2^{E}	1,375^{E}	7,625	9,000 (low?)
20 Qatif*	Saudi Arabia	1945	23.2	474	8,526	9,000 (low?)
21. Khurais	Saudi Arabia	1957	13.1	64	8,436	8,500 (?)
22. Zuluf**	Saudi Arabia	1965	24.3	152	8,348	8,500 (low?)
23. Raudhatain	Kuwait	1955	80.0^{E}	1,587^{E}	6,113	7,700 (?)
24. Sarir	Libya	1961	53.5^{E}	760^{E}	6,440	7,200 (?)
25. Hassi Messaoud	Algeria	1956	210.0	2,178	4,822	7,000 (?)
26. Shaybah	Saudi Arabia	1968	--	--	7,000	7,000 (low?)
27. Abu Sa'fah**	Saudi Arabia	1963	21.1	270	6,330	6,600
28. Asab	Abu Dhabi	1965	123.5^{E}	213^{E}	5,787	6,000 (?)
29. Bab	Abu Dhabi	1954	28.3^{E}	498^{E}	5,502	6,000 (?)
30. Ta-ch'ing	China	1959	285.0^{E}	1,725^{E}	4,275	6,000 (?)
31. East Texas	U.S.A.	1930	69.0	4,310	1,290	5,600 (low?)
32. Umm Shaif**	Abu Dhabi	1958	65.2	549	4,451	5,000 (?)
33. Wafra	Neutral Zone	1953	43.5	904	4,096	5,000 (?)
34. Samaria-Cunduacan	Mexico	1973	59.5	79	4,421	4,500 (?)
35. Zubair	Iraq	1948	73.0^{E}	776^{E}	3,724	4,500 (?)

Table 2.1--continued

Field	Country	Discovery Year	1975 Production	Cumulative Production	Reserves	Total Recovery
36. *Amal*	Libya	1959	26.6	510	3,740	4,250 (?)
37. *Nasser*	Libya	1959	46.9	1,740	2,460	4,200 (?)
38. *Gialo*	Libya	1961	70.2	1,181	2,819	4,000 (?)
39. *Khursaniyah*	Saudi Arabia	1956	13.9	486	3,514	4,000
40. *Minas*	Indonesia	1944	136.9	1,946	2,054	4,000 (?)
41. *Rag-e Safid*	Iran	1964	78.3	333	3,667	4,000 (?)
42. *Sabriyah*	Kuwait	1956	4.0[E]	208[E]	3,792	4,000 (?)
43. *Bibi Hakimeh*	Iran	1961	95.6	1,211	2,589	3,800 (?)
44. *Dukhan*	Qatar	1940	59.5	1,561	1,289	2,850
45. *Arlan*	U.S.S.R.	1955	100.0[E]	1,125[E]	1,675	2,800 (?)
46. *Wilmington Trend*[*]	U.S.A.	1922	70.9	1,973	827	2,800 (low?)
47. *Statfjord*[**]	Norway/U.K.	1974	--	--	2,570	2,570 (low?)
48. *Karanj*	Iran	1963	76.3	616	1,884	2,500 (low?)
49. *Paris*	Iran	1964	71.2	683	1,817	2,500 (?)
50. *Uzen*	U.S.S.R.	1961	115.0[E]	670[E]	1,830	2,500 (?)
51. *Balakhany-Sabunchi-Ramani*	U.S.S.R.	1896	5.0[E]	2,350[E]	50	2,400 (?)
52. *Haft Kel*	Iran	1928	10.8	1,682	718	2,400 (?)
53. *Malgobek-Vosnesensko-Aliyurt*	U.S.S.R.	1915	55.0[E]	2,000[E]	300	2,300 (?)
54. *Ust-Balyk*	U.S.S.R.	1961	110.0[E]	690[E]	1,610	2,300 (?)
55. *Idd el Shargi*[**]	Qatar	1960	5.0	150	1,950	2,100 (high?)
56. *Yates*	U.S.A.	1926	18.7	661	1,389	2,050 (low?)
57. *Bai Hassan*	Iraq	1953	14.0[E]	250[E]	1,750	2,000 (?)
58. *Brent*[**]	U.K.	1971	--	--	2,000	2,000 (?)
59. *Buzurgan*	Iraq	1969	--	--	2,000	2,000 (low?)
60. *Darius-Kharg*[*]	Iran	1961	42.4	479	1,521	2,000 (?)
61. *Minagish*	Kuwait	1959	21.0[E]	204[E]	1,796	2,000 (?)
62. *Mubarras*[**]	Abu Dhabi	1969	7.7	15	1,985	2,000 (?)
63. *Poza Rica*	Mexico	1930	19.9	1,093	907	2,000
64. *Sassan-Abu al Bukhoosh*[**]	Iran/Abu Dhabi	1966	86.2	443	1,557	2,000 (?)
65. *Tuymazy*	U.S.S.R.	1937	40.0[E]	1,660[E]	240	1,900 (?)
66. *Cactus-Nispero*	Mexico	1972	26.3	46	1,804	1,850 (?)
67. *Midway-Sunset*	U.S.A.	1901	37.3	1,265	565	1,830 (low?)
68. *Pembina*	Canada	1953	35.5	836	974	1,810 (?)
69. *Defa*	Libya	1960	48.3	394	1,406	1,800 (?)
70. *Forties*[**]	U.K.	1970	2.3	2	1,798	1,800 (low?)

Table 2.1--continued

Field	Country	Discovery Year	1975 Production	Cumulative Production	Reserves	Total Recovery
71. *Intisar D*	Libya	1967	66.4	591	1,209	1,800 (?)
72. *Wasson*	U.S.A.	1937	94.8	992	808	1,800 (low?)
73. *Mamontovo*	U.S.S.R.	1965	70.0[E]	200[E]	1,550	1,750 (?)
74. *Scurry*	U.S.A.	1948	79.1	1,051	699	1,750 (low?)
75. *Lamar***	Venezuela	1958	44.0[E]	745[E]	905	1,650 (low?)
76. *Ardeshir***	Iran	1974	--	--	1,600	1,600 (?)
77. *Kern River*	U.S.A.	1899	27.8	659	941	1,600 (low?)
78. *Mukhanovo*	U.S.S.R.	1945	55.0[E]	1,195[E]	355	1,550 (?)
79. *Abu Ghirab/Dehluran*	Iraq/Iran	1971	--	--	1,500	1,500 (low?)
80. *Bombay High***	India	1974	--	--	1,500	1,500 (?)
81. *El Morgan***	Egypt	1965	31.6	474	1,026	1,500 (?)
82. *Fateh***	Dubai	1966	52.9	279	1,221	1,500
83. *Mansuri*	Iran	1963	4.7	9	1,491	1,500 (low?)
84. *Nafoora-Augila*	Libya	1965	44.8	744	756	1,500 (low?)
85. *Slaughter-Levelland*	U.S.A.	1936	60.6	885	615	1,500 (low?)
86. *Souedie*	Syria	1959	49.3[E]	265[E]	1,235	1,500 (?)
87. *Umm Gudair*	Kuwait/ Neutral Zone	1962	29.7[E]	258[E]	1,242	1,500 (?)
88. *Kotur-Tepe*	U.S.S.R.	1956	70.0[E]	720[E]	740	1,460 (?)
89. *Sho-Vel-Tum*	U.S.A.	1915	32.8	1,034	391	1,425
90. *Panhandle*	U.S.A.	1910	11.5	1,295	115	1,410
91. *Elk Hills*	U.S.A.	1919	0.8	284	1,016	1,300 (low?)
92. *Novo-Elkhov*	U.S.S.R.	1955	60.0[E]	630[E]	670	1,300 (?)
93. *Masjid-e-Suleiman*	Iran	1908	4.7	1,195	105	1,300 (low?)
94. *Rimthan*	Saudi Arabia	1974	--	--	1,300	1,300 (low?)
95. *Bul Hanine***	Qatar	1970	50.7	172	1,078	1,250 (low?)
96. *Kingfish***	Australia	1967	74.2	314	936	1,250 (low?)
97. *Shkapovo*	U.S.S.R.	1953	40.0[E]	950[E]	300	1,250 (?)
98. *Sitio Grande*	Mexico	1972	26.8	43	1,207	1,250 (?)
99. *Sovetskoye*	U.S.S.R.	1962	50.0[E]	265	985	1,250 (?)
100. *Ekofisk***	Norway	1969	69.1	108	1,117	1,225 (low?)
101. *Bibi Eybat**	U.S.S.R.	1848	10.0[E]	1,110[E]	90	1,200 (?)
102. *Centro***	Venezuela	1957	40.0[E]	406[E]	794	1,200 (low?)
103. *Intisar A*	Libya	1967	21.3	524	676	1,200 (?)
104. *L-65*	Libya	1966	9.7[E]	122[E]	1,078	1,200 (?)
105. *Maydan-Mahzan***	Qatar	1963	39.2	439	761	1,200 (low?)

Table 2.1--continued

Field	Country	Discovery Year	1975 Production	Cumulative Production	Reserves	Total Recovery
106. *Ventura-Rincon**	U.S.A.	1916	15.7	990	210	1,200 (low?)
107. *Waha*	Libya	1960	36.5	560	640	1,200 (low?)
108. *Naranjos-Cerro Azul*	Mexico	1909	1.6	1,155	20	1,175 (?)
109. *Fateh Southwest***	Dubai	1970	40.0	118	982	1,100
110. *Leng-hu*	China	1958	3.0[E]	70[E]	1,030	1,100 (?)
111. *Ninian***	U.K.	1974	--	--	1,100	1,100 (?)
112. *Zarzaitine*	Algeria	1958	34.8	540	560	1,100 (?)
113. *Huntington Beach**	U.S.A.	1920	17.2	956	139	1,095 (low?)
114. *Abu Hadriya*	Saudi Arabia	1940	20.1	318	737	1,055 (low?)
115. *Neftyany Kamni**	U.S.S.R.	1949	28.0[E]	850[E]	200	1,050 (?)
116. *Southwest Ampa***	Brunei	1963	33.5	298	752	1,050
117. *Dammam*	Saudi Arabia	1938	6.7	545	500	1,045
118. *Boscan*	Venezuela	1946	18.0[E]	516[E]	524	1,040 (low?)
119. *Raguba*	Libya	1961	18.6	429	611	1,040 (high?)
120. *Harmaliyah*	Saudi Arabia	1971	15.6	75	950	1,025 (low?)
121. *Seria**	Brunei	1928	11.9	842	183	1,025
122. *Hawkins*	U.S.A.	1940	40.8	577	438	1,015
123. *Awali*	Bahrein	1932	22.3	591	409	1,000
124. *Bu Attifel*	Libya	1968	40.0	142	858	1,000 (low?)
125. *Fauqi*	Iraq	1974	--	--	1,000	1,000 (low?)
126. *Hamrin*	Iraq	1973	--	--	1,000	1,000 (low?)
127. *Jambur*	Iraq	1954	2.0[E]	67[E]	933	1,000 (?)
128. *Kupal*	Iran	1964	5.2	23	977	1,000 (low?)
129. *Pazanan*	Iran	1937	11.0	151	849	1,000 (low?)
130. *Ramadan***	Egypt	1974	13.5	14	986	1,000 (?)
131. *Swan Hills*	Canada	1957	43.2	387	613	1,000
132. *Urdaneta***	Venezuela	1955	--	2	998	1,000 (low?)
133. *Usin*	U.S.S.R.	1963	30.0[E]	52[E]	948	1,000 (?)
134. *Vozey*	U.S.S.R.	1972	--	--	1,000	1,000 (?)
135. *Ebano-Panuco*	Mexico	1901	2.7	932	43	975
136. *Fadhili*	Saudi Arabia	1949	5.4	162	798	960 (?)
137. *Greta-Tom O'Connor*	U.S.A.	1933	29.2	643	307	950
138. *La Paz*	Venezuela	1923	6.5	791	159	950 (?)
139. *Pravdinsk*	U.S.S.R.	1964	49.0[E]	201[E]	749	950 (?)
140. *Long Beach*	U.S.A.	1921	2.5	871	59	930 (low?)

Table 2.1--continued

Field	Country	Discovery Year	1975 Production	Cumulative Production	Reserves	Total Recovery
141. *Eunice Area*	U.S.A.	1927	12.7	655	245	900 (low?)
142. *Fahud*	Oman	1964	37.9	453	447	900 (?)
143. *Karatchouk*	Syria	1956	9.1[E]	41[E]	859	900 (?)
144. *Malongo West***	Angola	1969	34.5	134	766	900 (?)
145. *Oficina*	Venezuela	1937	10.6	664	236	900 (?)
146. *Quiriquire*	Venezuela	1928	4.9	733	167	900 (low?)
147. *Baku Archipelago**	U.S.S.R.	1963	34.0[E]	195[E]	665	860 (?)
148. *Goldsmith-Andector*	U.S.A.	1935	20.3	674	181	855 (low?)
149. *Halibut*	Australia	1967	46.6	304	546	850 (low?)
150. *Coalinga*	U.S.A.	1887	6.0	639	191	830 (low?)
151. *Redwater*	Canada	1948	35.5	576	229	805
152. *Bolshoye-Chernogor*	U.S.S.R.	1970	--	--	800	800 (?)
153. *Handil**	Indonesia	1974	1.6	2	798	800 (?)
154. *Jones Creek*	Nigeria	1967	29.0	211	589	800 (?)
155. *McElroy-Dune*	U.S.A.	1926	19.3	474	326	800 (low?)
156. *Moreni-Gura Ocnitei*	Rumania	1900	3.5[E]	758[E]	42[E]	800 (low?)
157. *Pokachev*	U.S.S.R.	1970	--	--	800	800 (?)
158. *Shushufindi*	Ecuador	1969	26.6	87	713	800 (?)
159. *Yarino*	U.S.S.R.	1956	40.0[E]	385[E]	415	800 (?)
160. *Rangely*	U.S.A.	1902	20.5	534	261	795 (low?)
161. *Hastings*	U.S.A.	1935	28.1	531	234	765
162. *Kuleshova*	U.S.S.R.	1958	40.0[E]	485[E]	275	760 (low?)
163. *Salt Creek*	U.S.A.	1889	10.0	560	200	760
164. *Oklahoma City*	U.S.A.	1928	1.9	736	19	755 (low?)
165. *Ab Teymur*	Iran	1968	--	--	750	750 (low?)
166. *El Borma*	Algeria/Tunisia	1964	23.9	274	476	750 (?)
167. *Fyzabad-Forest Reserve*	Trinidad and Tobago	1913	8.7	660	90	750 (?)
168. *Caillou Island**	U.S.A.	1930	14.1	529	206	735 (?)
169. *Conroe*	U.S.A.	1931	21.4	558	177	735 (low?)
170. *Karamai*	China	1955	7.8[E]	60[E]	670	730 (?)
171. *West Surgut*	U.S.S.R.	1962	40.0[E]	250	480	730 (?)
172. *Dahra-Hofra*	Libya	1959	7.0	445	255	700 (?)
173. *Guara-Guico*	Venezuela	1944	9.0[E]	588	112	700
174. *Imo River*	Nigeria	1959	22.6	280	420	700 (?)
175. *Infantas-La Cira*	Colombia	1918	7.4	626	74	700

Table 2.1--continued

Field	Country	Discovery Year	1975 Production	Cumulative Production	Reserves	Total Recovery
176. *July***	Egypt	1973	16.1	23	677	700 (?)
177. *Mene Grande*	Venezuela	1914	4.9	587	113	700 (low?)
178. *Piper***	U.K.	1973	--	--	700	700 (low?)
179. *Yokri*	Nigeria	1968	31.7	190	510	700 (?)
180. *Bradford*	U.S.A.	1871	2.1	661	19	680 (low?)
181. *Buena Vista*	U.S.A.	1910	3.4	619	31	650 (low?)
182. *Meren***	Nigeria	1965	27.9	224	426	650 (low?)
183. *Webster*	U.S.A.	1936	25.1	439	201	640
184. *Megion*	U.S.S.R.	1961	30.0^E	190^E	440	630 (?)
185. *Santa Fe Springs*	U.S.A.	1919	0.7	602	23	625 (low?)
186. *Bay Marchand Block 2**	U.S.A.	1949	28.3	461	149	610 (low?)
187. *Agan*	U.S.S.R.	1966	12.0^E	22^E	578	600 (?)
188. *Attaka***	Indonesia	1970	36.2	101	499	600 (?)
189. *Bahi*	Libya	1968	16.1	190	410	600 (low?)
190. *Brae***	U.K.	1975	--	--	600	600 (?)
191. *Duri*	Indonesia	1941	10.4	258	342	600 (?)
192. *Fedorov*	U.S.S.R.	1971	30.0^E	46^E	554	600 (?)
193. *Karachukhur-Zykh*	U.S.S.R.	1928	10.0^E	510^E	90	600 (?)
194. *Lung-nu*	China	1956	2.0^E	40^E	560	600 (?)
195. *Malongo North & South***	Angola	1966	16.8	149	451	600 (?)
196. *Mata*	Venezuela	1954	9.6^E	456	144	600 (?)
197. *Sacha*	Ecuador	1969	16.8	86	514	600 (?)
198. *North Varyegan*	U.S.S.R.	1971	--	--	600	600 (?)
199. *Yibal*	Oman	1962	34.5	149	451	600 (?)
200. *Dunlin***	U.K.	1973	--	--	585	585 (?)
201. *Nipa*	Venezuela	1945	6.9	442	133	575 (?)
202. *Bekasap*	Indonesia	1955	21.2	293	277	570 (?)
203. *Spraberry Trend*	U.S.A.	1952	18.2	418	152	570 (low?)
204. *Greater Jusepin*	Venezuela	1938	1.8^E	525^E	35	560 (low?)
205. *La Brea-Parinas*	Peru	1869	3.4	510	50	560 (?)
206. *Murchison***	U.K./Norway	1975	--	--	560	560 (?)
207. *Bassein North***	India	1975	--	--	550	550 (?)
208. *Thistle***	U.K.	1973	--	--	550	550 (?)
209. *Van*	U.S.A.	1928	15.8	435	115	550 (low?)
210. *Burbank*	U.S.A.	1920	3.4	507	33	540 (low?)

Table 2.1--continued

Field	Country	Discovery Year	1975 Production	Cumulative Production	Reserves	Total Recovery
211. *Smackover*	U.S.A.	1922	3.1	512	28	540 (?)
212. *Elk Basin*	U.S.A.	1915	8.0	449	86	535 (low?)
213. *San Ardo*	U.S.A.	1947	13.9	288	242	530 (low?)
214. *McArthur River***	U.S.A.	1965	40.9	294	231	525
215. *Ya-erh-hsia*	China	1957	4.0[E]	60[E]	465	525 (?)
216. *Soldado***	Trinidad and Tobago	1954	17.6	281	239	520 (?)
217. *Judy Creek*	Canada	1959	24.5	228	287	515
218. *South Sand Belt*	U.S.A.	1926	3.7	493	22	515 (low?)
219. *Borregos-Seeligson-T.C.B.*	U.S.A.	1938	4.8	478	32	510
220. *South Swan Hills*	Canada	1960	22.7	169	341	510
221. *Bangko*	Indonesia	1970	31.6	149	351	500 (?)
222. *Beryl***	U.K.	1972	--	--	500	500 (?)
223. *Binak*	Iran	1959	11.8	123	377	500 (?)
224. *Bomu*	Nigeria	1958	11.3	258	242	500 (?)
225. *Coalinga East Extension*	U.S.A.	1938	3.3	472	28	500
226. *Gassi Touil*	Algeria	1963	27.4	70	430	500 (?)
227. *Jana***	Saudi Arabia	1967	--	--	500	500 (low?)
228. *Kholmogor*	U.S.S.R.	1973	--	--	500	500 (?)
229. *Lab-e-Safid*	Iran	1969	6.6	18	482	500 (?)
230. *Luhais*	Iraq	1961	--	--	500	500 (?)
231. *Maharah***	Saudi Arabia	1973	--	--	500	500 (low?)
232. *Naft-i-Safid*	Iran	1935	13.2	280	220	500 (?)
233. *Nahr Umr*	Iraq	1948	--	--	500	500 (?)
234. *Okan*	Nigeria	1964	14.3	232	268	500 (?)
235. *Rio Nuevo*	Mexico	1975	0.3	--	500	500 (?)
236. *Saath al Raazboot***	Abu Dhabi	1969	--	--	500	500 (low?)
237. *Sahil*	Abu Dhabi	1970	--	--	500	500 (low?)
238. *Samah*	Libya	1961	11.3	248	252	500 (?)
239. *Shah*	Abu Dhabi	1966	--	--	500	500 (low?)
240. *Shurom*	Iran	1970	--	--	500	500 (?)
241. *Thompson (all)*	U.S.A.	1921	15.2	384	116	500
242. *Vat'yegan*	U.S.S.R.	1971	--	--	500	500 (?)
243. *Zhetybai*	U.S.S.R.	1960	20.0[E]	125[E]	375	500 (?)

Table 2.1--continued

Field	Country	Discovery Year	1975 Production	Cumulative Production	Reserves	Total Recovery
Combination Giants						
244. *Cushing*	U.S.A.	1912	2.7	466	24	490 (low?)
245. *Greater Matzen*	Austria	1949	10.9	363	127	490
246. *Weyburn-Midale*	Canada	1953	10.4	312	178	490
247. *Timbalier Bay**	U.S.A.	1939	11.6	410	75	485
248. *San Andres*	Mexico	1956	11.1	259	221	480 (?)
249. *South Cowden, Foster, Johnson*	U.S.A.	1932	16.2	331	139	470 (low?)
250. *North Cowden*	U.S.A.	1930	16.5	298	167	465 (low?)
251. *South Pass Block 24**	U.S.A.	1950	12.8	382	83	465
252. *Kettleman North Dome*	U.S.A.	1928	0.5	453	7	460 (low?)
253. *Rhourde El Baguel*	Algeria	1962	3.9	249	211	460 (?)
254. *Bonnie Glen*	Canada	1952	19.6	250	195	445
255. *Brea-Olinda-Sansinena*	U.S.A.	1884	3.9	395	45	440 (low?)
256. *Vacuum*	U.S.A.	1929	14.0	292	148	440 (low?)
257. *Seminole*	U.S.A.	1936	21.7	211	219	430 (low?)
258. *West Delta Block 30***	U.S.A.	1954	17.7	332	93	425 (low?)
259. *Eldfisk***	Norway	1970	--	--	420	420 (low?)
260. *Fullerton*	U.S.A.	1941	7.3	260	155	415 (low?)
261. *Arzanah*	Abu Dhabi	1973	--	--	400	400 (?)
262. *Valhall***	Norway	1975	--	--	400	400 (low?)
263. *South Pass Block 27***	U.S.A.	1954	9.7	268	107	375
264. *West Ranch*	U.S.A.	1938	13.8	295	80	375
265. *Leduc-Woodbend*	Canada	1947	6.9	335	30	365
266. *Dorra***	Neutral Zone	1967	--	--	340	340 (?)
267. *Malossa*	Italy	1973	0.5	1	324	325
268. *Grand Isle Block 43***	U.S.A.	1956	17.6	181	94	275
269. *Blinebry-Drinkard*	U.S.A.	1944	4.4	199	51	250
270. *Albuskjell***	Norway	1972	--	--	245	245 (?)
271. *Sand Hills*	U.S.A.	1930	4.6	190	45	235 (low?)
272. *South Kaybob*	Canada	1963	2.9	28	62	90

Table 2.1--continued

Field	Country	Discovery Year	1975 Production	Cumulative Production	Reserves	Total Recovery
Probable Giants						
South Belridge	U.S.A.	1911	9.3	206	214	420 (low?)
Caddo-Pine Island	U.S.A.	1906	3.1	320	25	345 (low?)
Eugene Island Block 330**	U.S.A.	1971	31.1	62	238	300 (low?)
Healdton	U.S.A.	1912	6.7	301	49	350 (low?)
Hondo**	U.S.A.	1969	--	--	95	95 (low?)
Howard Glasscock	U.S.A.	1925	7.0	317	113	430 (low?)
Loudon	U.S.A.	1938	2.3	361	24	385 (low?)
Oregon Basin	U.S.A.	1912	12.6	253	122	375 (low?)
Mara	Venezuela	1945	4.0	383	67	450 (low?)
Santa Rosa	Venezuela	1955	9.0^{E}	314^{E}	136	450 (low?)
Alwyn**	U.K.	1973	--	--	425	425 (?)
Andrew**	U.K.	1974	--	--	450	450 (?)
Claymore**	U.K.	1974	--	--	405	405 (low?)
North Cormorant**	U.K.	1972	--	--	400	400 (?)
Fulmar**	U.K.	1975	--	--	400	400 (?)
Magnus**	U.K.	1974	--	--	450	450 (?)
Belayem	Egypt	1955	11.5^{E}	275^{E}	200	475 (?)
Belayem Marine**	Egypt	1961	13.0^{E}	223^{E}	202	425 (?)
Grondin**	Gabon	1969	22.4	38	362	400 (low?)
Kokori	Nigeria	1961	18.6	192	258	450 (?)
Obagi	Nigeria	1966	19.3	106	294	400 (?)
Olomoro-Afiesere-Eriemu-Oweh	Nigeria	1961	21.3	298	302	600 (?)
Umm Addalkh**	Abu Dhabi	1968	--	--	300	300 (low?)
Rashid**	Dubai	1973	--	--		
Chesmeh Khush	Iran	1967	--	--	300	300 (low?)
Cyrus**	Iran	1962	10.8	59	341	400 (?)
Kilar Karim	Iran	1968	--	--	300	300 (low?)
Nowrouz**	Iran	1966	8.6	55	370	425 (?)
Rakhsh**	Iran	1969	10.9	63	237	300 (?)
Ramin	Iran	1966	--	--	250	250 (low?)
Qayarah	Iraq	1927	2.5^{E}	36^{E}		
Hout**	Neutral Zone	1963	3.0^{E}	39	161	200 (low?)
Natih	Oman	1963	17.7	212	188	400 (?)
Abu Jifan	Saudi Arabia	1973	--	--	279	279 (low?)
Lawhah**	Saudi Arabia	1975	--	--	276	276 (low?)
Mazalij	Saudi Arabia	1971	--	--	338	338 (low?)

SOURCES: See Appendix A.

NOTES: * = partly offshore.

** = offshore.

E = estimated.

(?) = uncertain estimate with a reasonable probability of at least a ± 10% change by 1988.

(low?) = uncertain estimate with a reasonable probability of at least a 10% increase by 1988.

(high?) = uncertain estimate with a reasonable probability of at least a 10% decrease by 1988.

also affected the composition of the list. All but three of the combination giant oil fields are in North America and Western Europe, the only two regions for which reasonable information on natural gas production and reserves is available on a systematic basis. If reliable information about natural gas was more readily available elsewhere, particularly for Southeast Asia, Nigeria, and the Soviet Union, the list of combination giant oil fields would be larger than shown here.

COMPARISON WITH OTHER COMPILATIONS

During the past decade, two other lists of giant fields have been compiled, one of which has been published in its entirety and the other in summary form.* The list in this report, with its inclusion of potential giant fields, is more comprehensive than either of the other two lists. The Halbouty-Klemme list contains 227 fields; the Mobil lists contain 233, 264, 268, and 299 fields. The list in this report also differs in composition from the other two, excluding several fields that they include either for methodological reasons, a lack of confirming evidence from other sources, or the presence of sufficient information to indicate that the field in question does not merit giant status.

Thirty-seven of the fields in the Halbouty-Klemme list were either downgraded to potential giants or dropped from the list altogether. Seven (*Belayem, Belayem Marine, Claymore, Eugene Island Block 330, Hout, Natih,* and *Obagi*) are considered to be probable giants in this report.

---

*The former is the one by Halbouty et al., "World's Giant Oil and Gas Fields." This list was subsequently published with minor additions in Klemme, "The Giants and the Supergiants," and has since been updated in Klemme, "Giant Fields Contain Less Than 1% of World's Fields But 75% of Reserves." The latter lists, compiled and subsequently updated by the geologic staff of Mobil Oil Company, have been summarized and analyzed, but never listed, in several articles, including Moody, "Distribution and Geological Characteristics of Giant Oil Fields"; Moody and Esser, "An Estimate of the World's Recoverable Crude Oil Resource"; Holmgren, Moody, and Emmerich, "The Structural Settings for Giant Oil and Gas Fields"; and Meyerhoff, "Economic Impact and Geopolitical Implications of Giant Petroleum Fields." These lists, which are undocumented, appear to be based primarily on the original list by Halbouty et al. and the annual summaries of production by field published in the *Oil and Gas Journal*. The two will be subsequently referred to as the Halbouty-Klemme list and the Mobil lists.

Fourteen are listed here as possible giants (*Auca, Bushgan, Cheleken, West Ekofisk, Emeraude, Glynsko-Rozybyshev, Lawrence* (part of *Old Illinois*), *Loango, Oktyabr'skoye, Priluki, Rostam, Salym, Shengli,* and *Starogrosni*). Five were dropped because they did not meet the field definition used here (*Golden Trend, Lima-Indiana,* and *Greater Seminole* in the United States; *Rainbow* in Canada; and *Comodoro Rivadavia* in Argentina). Six are either combination giant gas fields or smaller gas fields (*Barqan,* Saudi Arabia; *Barracouta* and *Marlin,* Australia; *Bastian Bay* and *Bayou Sale*, United States; and *Korobki,* Soviet Union). Four were dropped on the basis of subsequent information indicating that the fields were only erstwhile giants (*Amal,* Egypt; *Arenque,* Mexico; *South Cheremshanka*, Soviet Union; and *West Delta Block 73*, United States). No mention of another (*Diamond Trend,* United States) could be found, despite the thorough documentation available on Texas oil fields (this field may be the one referred to in this report as the *South Sand Belt*).

Some of the few detectable differences between the list in this report and the Mobil lists repeat the differences between the list of this report and the Halbouty-Klemme list, particularly the differences in field definition. There are several differences of judgment about fields in South America and Asia-Oceania that are unique. The Mobil lists indicate up to three more giant fields in Argentina, none of which seem at all likely in view of what is known about the fields in each province and the availability of production data by province (Mobil may have confused the province data in the *Oil and Gas Journal* with field data). Mobil lists two to three more giant fields in the Oriente Basin of Colombia and Ecuador (most likely *Orito* and *Lago Agrio*, two large but most likely non-giant fields). One more giant in coastal Peru is indicated (most likely the Lima coastal fields, which together have a total recovery of about 450 million barrels and could prove to be of giant status). Up to two more giant oil fields are indicated for the Gippsland Basin in Australia (possibly *Marlin* and *Snapper*, both of which are combination gas giant fields). One giant field is indicated for Assam in India (possibly either *Digboi* or *Nahorkatiya*, both of which are likely to be larger than 100 million barrels each but are clearly not giants).* The number of known and potential giants listed in this report in the various regions of North America, Western Europe, the Soviet Union, Africa, and the Middle East are equal or greater than the number of giant fields in those regions shown in the Mobil summaries.

Given the limitations of the available data, particularly on China and on discovered but nonproducing fields in the North Sea, the Soviet Union, and the Middle East, and subject to the methodological guidelines laid out in this section, the list of giant fields (including potential giants) in this report provides a comprehensive compilation of the giant oil fields of the world with reasonably accurate indications of their known contributions to world oil resources.

---

*For information on these fields, see E. N. Tiratsoo, *Oilfields of the World*, Scientific Press, Beaconsfield, England, 1973.

III. THE DISTRIBUTION OF WORLD OIL RESOURCES

Popular discussions of known and potential conventional oil resources typically focus on single aggregate estimates of ultimate resources. As useful as such aggregates are in focusing attention on some fundamental characteristics of the global energy picture, by themselves they do not convey much information and they can be highly misleading. A reasonable understanding of the world oil outlook is not possible without some knowledge of the composition of these aggregates.

This section describes several aspects of the composition of known conventional world crude oil resources. The distribution of known recoverable resources by size of field, by country, and by province is outlined and briefly analyzed. In each case, the significance of giant fields is emphasized.

As of the end of 1975, 335.1 billion barrels of crude oil had already been produced, leaving 676.4 billion barrels of proved and probable reserves. Together these constitute a known recoverable crude oil resource of 1011.5 billion barrels.

These estimates of reserves and total recovery are conservative. They only include those recoverable resources that could be clearly identified from known and planned investment in production facilities. As Table 2.1, Section II, and the notes to the tables of Appendix A indicate, many of the estimates of total recovery from the known giant fields are lower than likely ultimate recovery from these fields. For many of the potential giant fields, no estimate of total recovery was even available. As will be discussed in Section V, ultimate recovery from known fields is likely to be considerably higher than indicated in Tables 2.1 and 3.1 (below), given the growth in reserves resulting from additional investments in recovery. These additions are not dealt with in what follows, the discussion being limited to the distribution and discovery patterns of only the known recoverable resources.

The estimate of known total recovery developed in this report is compared with four other recent, widely used estimates in Table 3.1.

Table 3.1

COMPARATIVE ESTIMATES OF KNOWN TOTAL RECOVERY OF CRUDE OIL BY REGION

(In billions of barrels)

Region	*International Petroleum Encyclopedia* (as of 1/1/76)	*World Oil* (as of 1/1/76)	Moody and Esser (as of 1/1/74)	Meyerhoff (as of 1/1/75)	This Report (as of 1/1/76)
U.S. and Canada	156.4	155.7	168	171	163.2
Latin America	81.5	75.1	83	88	85.0
Western Europe	28.2	21.6	24	29	24.6(+)
U.S.S.R., Eastern Europe	134.6	112.9	140[a]	143[a]	102.4(+)
North Africa	53.5	52.9	46	47[a]	52.9(+)
Central and Southern Africa	32.1	21.3	31	31[a]	22.7
Middle East	453.6	425.6	499	538	509.9(+)
China	23.0	20.2	10[a]	20[a]	23.0(+)
Asia-Oceania	29.9	28.9	36	38	27.8
World	992.8	914.2	1,037	1,105	1,011.5(+)

[a]Estimated division.

SOURCES:

1. *International Petroleum Encyclopedia, 1976*, Vol. 9, Petroleum Publishing Company, Tulsa, Okla., 1976, pp. 302-303.
2. *World Oil*, Vol. 185, No. 3, August 15, 1976, p. 44.
3. J. D. Moody and R. W. Esser, "An Estimate of the World's Recoverable Crude Oil Resource," *Proceedings, Ninth World Petroleum Congress*, Vol. 3 (Exploration and Transportation), Applied Science Publishers, London, 1975, p. 13.
4. A. A. Meyerhoff. "Economic Impact and Geopolitical Implications of Giant Petroleum Fields," *American Scientist*, Vol. 64, No. 5, September-October 1976, p. 540. Estimated in part using data from Moody and Esser and from D. A. Holmgren, J. D. Moody, and H. H. Emmerich, "The Structural Settings for Giant Oil and Gas Fields," *Proceedings, Ninth World Petroleum Congress*, Vol. 3 (Exploration and Transportation), Applied Science Publishers, London, 1975, p. 46.

The estimate is disaggregated by nine different regions in order to permit more meaningful comparisons. Because some of the differences among the five estimates are more apparent than real, comparisons among them need to be made carefully.

There are two apparent but not real differences among the estimates. The first is a conceptual one. The estimates of the *International Petroleum Encyclopedia* and *World Oil* include only proved reserves or at most proved plus some probable reserves. The estimates of Moody and Esser, Meyerhoff, and this report include both proved and probable reserves. However, the definition of probable reserves used in each of these three estimates is not likely to be the same. These conceptual differences account for practically all of the differences among the five estimates for the United States and Canada and for the Middle East. The second difference is the extent to which reserves in recent discoveries are included. This factor accounts for nearly all the differences among the estimates for Mexico (included here in Latin America), the North Sea (included here in Western Europe), and China.

The principal real difference among the five estimates is the estimate of known total recovery in the Soviet Union. The difference between the highest (Meyerhoff) and lowest (this report) is roughly 40 billion barrels. Nearly all of this difference is in giant fields. As the notes to Table A.36 indicate, in view of recent Soviet statements about declining reserves and reserve-production ratios and in view of the recent production declines or projected production declines for several of the major producing regions and giant fields in the Soviet Union, the high estimates with their implied nationwide reserve-production ratios of 20:1 or more are simply unsupportable from the available evidence.

The other major differences are Nigeria (in Central and Southern Africa) and Indonesia (Asia-Oceania). A difference of up to 7 billion barrels in Nigeria could be primarily a conceptual one; the low estimates may exclude reserve additions from secondary recovery projects not yet installed or planned that are included in the higher estimates. The information compiled for this report was insufficient to determine whether the difference was real or only conceptual. The difference of up to 5 billion barrels for Indonesia is clearly a real one. Here, too, the lower estimate appears to be better supported. Because many of the major fields are now at or near peak production and appear to have reserve-production ratios anywhere between 8:1 and 12:1, reserve estimates suggesting countrywide reserve-production ratios higher than the 19:1 as of the end of 1975 implied in Table A.69 are highly dubious given present and planned investment. However, improved economic incentives for operators could increase total recovery in Indonesia to the levels indicated by the higher estimates. The difference among estimates for Libya (in North Africa) of up to 4 billion barrels may not be significant, for the degree of uncertainty about known recoverable oil resources in this country permits differences up to 10%. The other differences among the five estimates are all small (generally 1 billion barrels or less per country) and are basically insignificant in terms of the overall estimate.

THE DISTRIBUTION OF WORLD OIL RESOURCES BY SIZE OF FIELD

Known recoverable world crude oil resources are heavily concentrated in giant fields. As Table 3.2 indicates, at least 76.7% of known recoverable oil resources as of the end of 1975 are in the 272 known giant fields. Another 4.3% are in the potential giant fields, and an additional 7.4 percent are in large non-giant fields for a total of 88.4 percent of the total recoverable resources in 908+ giant and large non-giant fields.*

The actual distribution is probably slightly different from the one indicated in Table 3.2. With more information, all of the potential giant fields could be allocated to either the small and combination giant category or the large non-giant category. The amounts in large

---

*By including potential giant fields in its scope, this report indicates an even greater importance for giant fields than have previous studies. The recent consensus is that 75% to 76% of known resources are in giant fields (a finding replicated in this report for known giants). See Holmgren, Moody, and Emmerich, "The Structural Settings for Giant Oil and Gas Fields," p. 46; Meyerhoff, "Economic Impact and Geopolitical Implications of Giant Petroleum Fields," p. 540; and Klemme, "Giant Fields Contain Less Than 1% of World's Fields But 75% of Reserves," p. 164.

Table 3.2

THE DISTRIBUTION OF KNOWN RECOVERABLE OIL RESOURCES OF THE WORLD BY FIELD SIZE AS OF DECEMBER 31, 1975

Field Size (In millions of barrels)	Number	Amount (In billions of barrels)	Percent of Total
Super-giants (5000 +)[a]	33	512.6	50.7
Large giants (2000-5000)[a]	31	89.3	8.8
Medium giants (1000-2000)[a]	70	90.2	8.9
Small and combination giants (500-1000)	138	84.0	8.3
Potential giants[a,b]	153(+)	43.1(+)	4.3(+)
Large non-giants (100-500)[b,c]	483(+)	75.2(+)	7.4(+)
Other non-giants (less than 100)[c]		117.1(-)	11.6(-)
Total[c]		1,011.5(+)	

[a]Because no estimates of total recovery are available for 30 of the potential giants and some estimates available for the others are likely to be substantially understated, the actual amount in potential giant fields could easily be 25%-50% greater than that shown here.

[b]Because some of the potential giants are likely to prove to be merely large non-giant fields, the amount in large non-giants is somewhat greater than indicated here.

[c]No information on large non-giant fields in China, Eastern Europe, and the Soviet Union could be obtained for this report. The amounts in these fields are therefore included in the amount in other non-giant fields. Transferring them to the large non-giant category would probably increase the amount in large non-giants by 15 to 20 billion barrels and decrease the amount in other non-giants by a similar amount.

SOURCES: See Appendix A, particularly Tables A.1, A.6, A.16, A.28, A.37, A.48, and A.63.

non-giant fields in China, Eastern Europe, and the Soviet Union, included here with other non-giants for lack of information, could be put into their proper category. Allowances for some reserve growth would shift some amounts from several of the categories into the next higher category. With these adjustments, including the exclusion of the potential giant category, super-giant fields would still contain approximately 50% of known recoverable resources as of the end of 1975, and each of the other five categories would contain 9% to 11% of known recoverable resources.

Regardless of the specific distribution used, it is clear that most of the world's oil has been found in a very small number of fields. More than half is in the 33 recognized super-giants. The four largest alone have over one-fifth (217 billion barrels). The 17 fields with 10 billion barrels or more each have nearly two-fifths of the total (398.9 billion barrels). Over two-thirds are in the 134 medium, large, and super-giant fields. This distribution of known recoverable oil resources by size of field indicates unmistakably that the increase in known recoverable world oil resources to their present level has been associated with the discovery and development of giant and particularly super-giant oil fields.

THE DISTRIBUTION OF WORLD OIL RESOURCES BY COUNTRY

World oil resources are also concentrated within a few countries. Table 3.3 indicates the distribution of identified recoverable oil (including cumulative production) by country as of the end of 1975. The countries are ranked in order of size of identified recoverable resources (from Appendix A) and grouped by identified plus additional probable resources from fields for which no estimates were available. (Some may not appear to qualify for a particular group because these additional reserves are not included in the totals shown. However, the additional reserves indicated by the plus sign were deemed sufficient to group them in the next higher category.) The number of known and potential giant fields in each country and the amount of recoverable oil in each type of field are shown as well. Because Table 3.3 includes cumulative production as well as proved and probable reserves,

Table 3.3

THE DISTRIBUTION OF KNOWN RECOVERABLE OIL RESOURCES AND GIANT OIL FIELDS BY COUNTRY AS OF DECEMBER 31, 1975
(In billions of barrels)

Country	Recoverable Oil	Known Giants		Potential Giants	
		Number	Amount	Number	Amount
Over 100 Billion Barrels					
Saudi Arabia	200.2(+)	19	199.0	11	1.0(+)
United States	148.0	59	61.3	28	8.1(+)
Iran	99.0(+)	23	93.6	26	3.8(+)
Soviet Union	96.0(+)	33	62.6	13	2.3(+)
50-100 Billion Barrels					
Kuwait	86.7(+)	5	86.7	2	N.A.
Venezuela	50.8	13	42.8	10	3.6
Iraq	49.8(+)	11	49.3	4	N.A.
25-50 Billion Barrels					
Abu Dhabi	42.8	11	42.4	5	0.4(+)
Libya	35.0	15	32.2	1	N.A.
China	23.0(+)	5	9.0	8(+)	10.0(?)
10-25 Billion Barrels					
Nigeria	18.0	6	3.9	5	2.2
Mexico	16.5(+)	8	12.7	1	0.3
Indonesia	15.5	6	7.1	7	2.1
Canada	15.2	9	6.0	--	--
United Kingdom	14.1	10	8.6	9	3.5
Neutral Zone	13.6	4	13.3	2	0.2(+)
Algeria	12.0	5	9.3	3	0.9
5-10 Billion Barrels					
Qatar	7.5	4	7.4	1	0.1(+)
Norway	5.5	6	4.6	2	0.4
Argentina	5.2	--	--	1	N.A.
Egypt	5.0	3	3.2	2	0.9
Rumania	5.0	1	0.8	3	N.A.
1-5 Billion Barrels					
India	3.6	2	2.1	--	--
Australia	3.3	2	2.1	2	0.6
Oman	3.2	2	1.5	1	0.4
Brazil	2.8	--	--	1	0.4
Syria	2.8(+)	2	2.4	2	0.2(+)
Brunei	2.7	2	2.1	1	0.3

Table 3.3--continued

Country	Recoverable Oil	Known Giants		Potential Giants	
		Number	Amount	Number	Amount
Dubai	2.6(+)	2	2.6	1	N.A.
Colombia	2.6	1	0.7	--	--
Trinidad and Tobago	2.5	2	1.3	--	--
Ecuador	2.1	2	1.4	1	0.3
Gabon	1.9	--	--	1	0.4
Angola	1.8	2	1.5	--	--
Malaysia	1.8	--	--	2	0.7
Peru	1.7	1	0.6	--	--
Federal Republic of Germany	1.5	--	--	1	0.1
Bahrein	1.0	1	1.0	--	--
0.1-1 Billion Barrels					
Tunisia	0.95	1	0.5	1	0.3
Congo	0.73	--	--	2	0.7
Austria	0.72	1	0.5	--	--
Yugoslavia	0.70	--	--	--	--
Italy	0.63	1	0.3	1	N.A.
Hungary	0.55	--	--	--	--
Turkey	0.55	--	--	--	--
Burma	0.45	--	--	--	--
France	0.43	--	--	--	--
Netherlands	0.43	--	--	1	0.2
Bolivia	0.42	--	--	--	--
Chile	0.40	--	--	--	--
Poland	0.39	--	--	--	--
Albania	0.32	--	--	--	--
Spain	0.30	--	--	--	--
Zaire	0.30	--	--	--	--
Denmark	0.20	--	--	--	--
Sharjah	0.20	--	--	--	--
Japan	0.18	--	--	--	--
Greece	0.15	--	--	--	--
New Zealand	0.13	--	--	--	--
Pakistan	0.13	--	--	--	--
Less Than 0.1 Billion Barrels					
Bulgaria	0.06	--	--	--	--
Czechoslovakia	0.05	--	--	--	--
East Germany	0.04	--	--	--	--
Guatemala	0.03	--	--	--	--
Taiwan	0.03	--	--	--	--
Israel	0.02	--	--	--	--
Morocco	0.02	--	--	--	--

SOURCES: See Appendix A.

its ranking of countries would differ from one based on reserves only, as a comparison between it and the estimates of reserves by country provided in Appendix A indicates.

Four countries--Saudi Arabia, the United States, the Soviet Union, and Iran--have among them more than half of the oil resources of the world that have been discovered and made recoverable since the beginning of the oil industry. The seven largest, including Kuwait, Venezuela, and Iraq, have nearly three-fourths of the known resource. The 17 largest (counting the Kuwaiti-Saudi Neutral Zone as a separate entity) with 10 billion barrels or more each (i.e., 1% or more of total known world oil resources) have 93% of the known resource. Fifty other countries account for the remaining 7%.

The distribution of oil resources by country is essentially identical with the distribution of giant oil fields. The four countries with the largest known recoverable resources have approximately half of the known giant fields among them (137 out of 272) and 19 of the 33 super-giants. Each is likely ultimately to have more than 25 giant fields. All of the 17 countries with 10 billion barrels or more have at least five known or potential giant fields each. Only one country (Norway) with less than 10 billion barrels has five or more known giants, and it could easily move into that category with more exploration and development in the North Sea. Except for Argentina and Rumania, which are both marginally in the category of 5 billion barrels or more and for which knowledge of field size distribution is hampered by a lack of information, all the 22 countries with 5 billion barrels or more have five or more known and potential giant fields.

Because of the concentration of total world oil resources in giant fields, the correlation between the distribution of giant fields and the distribution of resources among the major countries is not at all surprising. What is surprising is that this correlation extends throughout the list. Every country with at least 1 billion barrels of known recoverable oil resources has at least part of one known or potential giant oil field. (The Federal Republic of Germany, with a small proportion of a marginally potential giant, admittedly barely meets this criterion, but it is also near the bottom of the list. All of the other countries have known giants or fairly large potential giants.) In fact, all but three of the countries with no known or potential giant fields have a total recovery of less than 500 million barrels, the minimum level of recoverable oil for a giant oil field. Moreover, two of these three, Turkey and Hungary, are just at or barely above this level.

The concentration of world oil resources in those countries with giant oil fields does not mean that the discovery of giant oil fields is essential for the discovery of significant amounts of oil. As the estimates for the United States, the Soviet Union, Nigeria, and Canada in Table 3.3 indicate, significant amounts have been found in some countries in non-giant fields. The differences in the proportion of total resources found in giant oil fields depends primarily on differences in the distribution of province types and province sizes among countries. Some types of provinces are more likely to have the bulk of their recoverable resources in moderate to large rather than giant fields, tertiary delta provinces being the most prominent example.[*] In general, however, significant amounts of oil are closely associated with giant oil fields.

THE DISTRIBUTION OF WORLD OIL RESOURCES BY PROVINCE

Provinces are the basic regional geologic units for petroleum exploration and production.[†] A producing petroleum province is defined as a set of oil and gas fields that are geographically contiguous and that occur in a similar or related geologic environment.[‡] Provinces can

---

[*] For a description and analysis of basin types and the distribution of giant fields among them, see Klemme's "The Giants and the Supergiants" and "World Oil and Gas Reserves from Analysis of Giant Fields and Petroleum Basins (Provinces)." A related analysis, based on the concept of petroleum zone, is given in G. Gess and C. Bois, "Study of Petroleum Zones: A Contribution to the Appraisal of Hydrocarbon Resources," in Meyer (ed.), *The Future Supply of Nature-Made Petroleum and Gas*, pp. 155-178.

[†] The term "basin" is often used for "province," but the latter has gradually become the preferred term because of its more inclusive character.

[‡] Levorsen, *Geology of Petroleum*, p. 31; and Hobson and Tiratsoo, *Introduction to Petroleum Geology*, p. 6.

vary greatly in size, referring both to massive areas such as the West Siberian Basin (Soviet Union) or the Arabian-Iranian Basin (Middle East) and to quite small areas such as the Santa Maria Basin (United States) or the Middle Magdalena Basin (Colombia).

Known recoverable crude oil resources are heavily concentrated within a few producing provinces. Table 3.4 lists those provinces with a known recovery of 10 billion barrels or more or those that with further discoveries, development, and implementation of secondary and tertiary recovery are likely to reach 10 billion barrels or more. These provinces will be defined here as the *major provinces*. Current estimated total recovery is shown for each province (from Appendix A and its sources), together with the number of known and potential giant fields in each, current total recovery in each class of giants, and the year in which the first and latest giant field in each was discovered.

Cumulative production and identified reserves in the 22 major provinces listed in Table 3.4 total approximately 866 billion barrels, 86% of the oil discovered and made recoverable in the world to the end of 1975. This predominance stems from the fact that they contain all of the super-giant fields and 227 of the 272 known giants with 738.4 out of the 776.1 billion barrels in known giants. They also have 112+ of the 153+ potential giants with 23.2+ of the 43.1+ billion barrels in potential giant fields.

The major provinces are only a small proportion of the known provinces of the world. There are at least 400 provinces, about 275 of which have been subject to at least moderate exploratory efforts. Of the 125 that have only been lightly explored or have not been explored at all, 70 are either in polar regions or are deep ocean basins. Of the 275 that have been explored to at least a moderate degree, production or significant discoveries have occurred in at least 125. Of these 125, known giant oil fields have been discovered in at least 51 and giant gas fields in 33 for a total of 68 provinces with at least one known giant oil or gas field (counting the provinces with both only once). Another 17 of the remaining 57 provinces have potential

Table 3.4

THE DISTRIBUTION OF KNOWN RECOVERABLE OIL RESOURCES AND GIANT OIL FIELDS BY MAJOR PROVINCE AS OF DECEMBER 31, 1975

(In billions of barrels)

Province (Country)	Province Total	Known Giants		Potential Giants		Discovery Year	
		Number	Amount	Number	Amount	First Giant	Latest Giant
Alberta (Canada)	12.0[E]	8	5.5	--	--	1947	1963
Reforma (Mexico)	8.1	4	8.1	--	--	1972	1975(1977)[h]
North Slope (U.S.)	9.6	1	9.6	1	N.A.	1969	1975
San Joaquin (U.S.)	10.7	7	7.2	4	1.0	1899	1938
Los Angeles (U.S.)	8.2	5	5.9	--	--	1884	1922(1976)[h]
Permian[a] (U.S.)	25.6	16	13.4	7	2.3	1923	1952
East Texas[b] (U.S.)	9.2	3	7.2	--	--	1928	1940
Gulf Coast[c] (U.S.)	14.1	7	4.5	1	0.3	1921	1938
Mississippi Delta[d] (U.S.)	16.3	7	3.4	3	0.8	1930	1971
Maracaibo (Colombia, Venezuela)	40.4[E]	7	38.5	2	0.9	1914	1958
Maturin (Trinidad, Venezuela)	12.9[E]	8	5.5	7	2.5	1913	1958
North Sea (Denmark, Norway, U.K.)	19.8	14	13.3	11	3.9	1969	1975(1976)[h]
Volga-Ural[e] (USSR)	35.0[E]	8	22.8	--	--	1937	1958
West Siberian (USSR)	31.0[E]	14	25.0	6	1.1+	1961	1973
North Caucasus-Kopet Dag (USSR)	21.0[E]	9	12.9	4	0.9+	1848	1963
Erg Oriental (Algeria, Tunisia)	9.8[E]	4	8.7	2	0.6	1956	1965
Sirte (Libya)	34.0	15	32.2	1	N.A.	1959	1968
Niger Delta (Nigeria)	18.0	6	3.9	5	2.2	1958	1968
Arabian-Iranian[g] (Middle East)	509.9+	79	499.1	50	6.1+	1908	1975(1976)[h]
Sung-liao (China)	7.0+[E]	1	6.0	2(?)	N.A.	1959	?
North China (China)	5.0+[E]	--	--	4+(?)	N.A.	1962	?
Central Sumatra (Indonesia)	8.5[E]	4	5.7	2	0.6	1941	1970

[E]Estimated.

[N.A.]Not Available.

[a]Defined here as Southeast New Mexico and Texas R.R.C. Districts 7C, 8, and 8A.

[b]Defined here as Texas R.R.C. Districts 5 and 6.

[c]Defined here as Texas R.R.C. Districts 1, 2, 3, and 4.

[d]Defined here as South Louisiana.

[e]Defined here as Bashkiria, Kuibyshev, Lower Volga, Orenberg, Perm, Saratov, Tataria, and Udmurt.

[f]Defined here as North Caucasus, Baku, Turkmenia, and Mangyshlak.

[g]Defined here as all of the Middle East, excluding Israel.

[h]Post-1975 discoveries are in parentheses.

SOURCES: The province names are taken with some adaptations from M. T. Halbouty et al., "World's Giant Oil and Gas Fields, Geologic Factors Affecting Their Formation, and Basin Classification," M. T. Halbouty, ed., *Geology of Giant Petroleum Fields*, American Association of Petroleum Geologists Memoir 14, Tulsa, Okla., 1970, pp. 510-528, and H. D. Klemme, "The Giants and the Super-giants," *Oil and Gas Journal*, Vol. 69, Nos. 9, 10, and 11, March 1, 8, and 15, 1971. The estimates of recovery by province and fields, the number of giant fields within them, and the discovery years are from Appendix A and its sources.

giant oil fields.* Of the 30 or more provinces with giant fields other than the major provinces, only 13 have a known recovery of 3 billion barrels or more (Amarillo-Anadarko-Ardmore, Appalachian, Arkla, Bend-Red River, Illinois, Oklahoma Platform, Ventura, Tampico Embayment, Oriente, Pre-Carpathian, Suez, Gippsland, and North Borneo).†

The Arabian-Iranian Basin is clearly unique among all the provinces of the world.‡ It contains more than half of the known recoverable oil in the world. No other province has even 10% of its conventional oil resources, although extensive development of the oil (tar) sand deposits in the Alberta and Maturin basins would eventually push them above that level. The West Siberian Basin, with more than 700 trillion cubic feet of natural gas reserves, is the only province that even begins to have comparable conventionally recoverable petroleum and natural gas resources, and it is only one-fourth the size of the Arabian-Iranian Basin when measured in these terms. Only five provinces--the Permian, Maracaibo, Volga-Ural, West Siberian, and Sirte--have even 5% to 10% of the known oil resources of the Arabian-Iranian Basin. (The Niger Delta, North Caucasus-Kopet Dag, North Sea, and Reforma provinces may also eventually reach that level.) This predominance stems primarily from the concentration of super-giant oil fields in the Middle East. It has 25 of the 33 recognized super-giants. No other province has more than one (although Western Siberia has several super-giant gas fields).

---

*The number and names of provinces are derived with minor amendments from Klemme's "World Oil and Gas Reserves from Analysis of Giant Fields and Petroleum Basins (Provinces)" and "The Giants and the Super-giants."

†The various Rocky Mountain provinces (Powder River, Big Horn, Wind River, Green River, Overthrust Belt, Uinta-Piceance, Paradox, and San Juan) would form a major province if considered as one (which they were considered to be before the Laramide orogeny). They are considered separately here, following standard usage in the United States. See R. F. Meyer, "Geologic Provinces Code Map for Computer Use," *American Association of Petroleum Geologists Bulletin*, Vol. 54, No. 7, July 1970, pp. 1301-1305.

‡Other authors have subdivided the Arabian-Iranian Basin into several provinces. Although it contains several distinct petroleum zones, it has sufficient geological continuity to be considered as a single province. See Z. R. Beydoun and H. V. Dunnington, *The Petroleum Geology and Resources of the Middle East*, Scientific Press, Beaconsfield, England, 1975.

One other aspect of the concentration of world oil resources is noteworthy. The concentration of world oil resources in an arc from Algeria through the Middle East and the central Soviet Union to the Arctic Ocean has been emphasized by several authors.* An even more striking concentration is that of the "ring of oil" (Fig. 3.1). This oval band, imposed on a map showing the postulated position of the continental plates 180 million years ago, is roughly 800 to 1000 miles wide and contains nearly 85% of the known petroleum resources of the world (approximately 845 of 1012 billion barrels). Within it are found 14 of the 22 major provinces and all of the seven largest provinces. It also contains 210 of the 272 known giants with 706.1 of the 776.1 billion barrels in these fields. Most of the major nonconventional oil deposits of the world are also within the ring, notably the oil (tar) sand deposits of the Alberta and Maturin basins† and the oil shale deposits of the western Rocky Mountain basins. Moreover, there are few lengthy gaps with no petroleum production within the entire band (assuming the plate positions of Fig. 3.1). Only the nonsedimentary shield (and hence unprospective) areas of north central South America and West Africa and the relatively unexplored Arctic Islands have no established production. This continuity of production exists despite considerable differences in province types and in the geologic ages of productive formations among the producing provinces within the ring. No complete explanation for this phenomenon is yet available. However, the concentration of total resources within it is so great and the continuity of production is so remarkable that it poses a most interesting problem for geologic speculation.

---

*Halbouty et al., "World's Giant Oil and Gas Fields," p. 528; and Meyerhoff, "Economic Impact and Geopolitical Implications of Giant Petroleum Fields," pp. 539-540.

†Demaison, "Tar Sands and Supergiant Oil Fields"; and P. H. Phizackerley and L. O. Scott, "Major Tar Sand Deposits of the World," *Proceedings, Seventh World Petroleum Congress*, Vol. 3 (Drilling and Production), Elsevier, Barking, England, 1967, pp. 551-571.

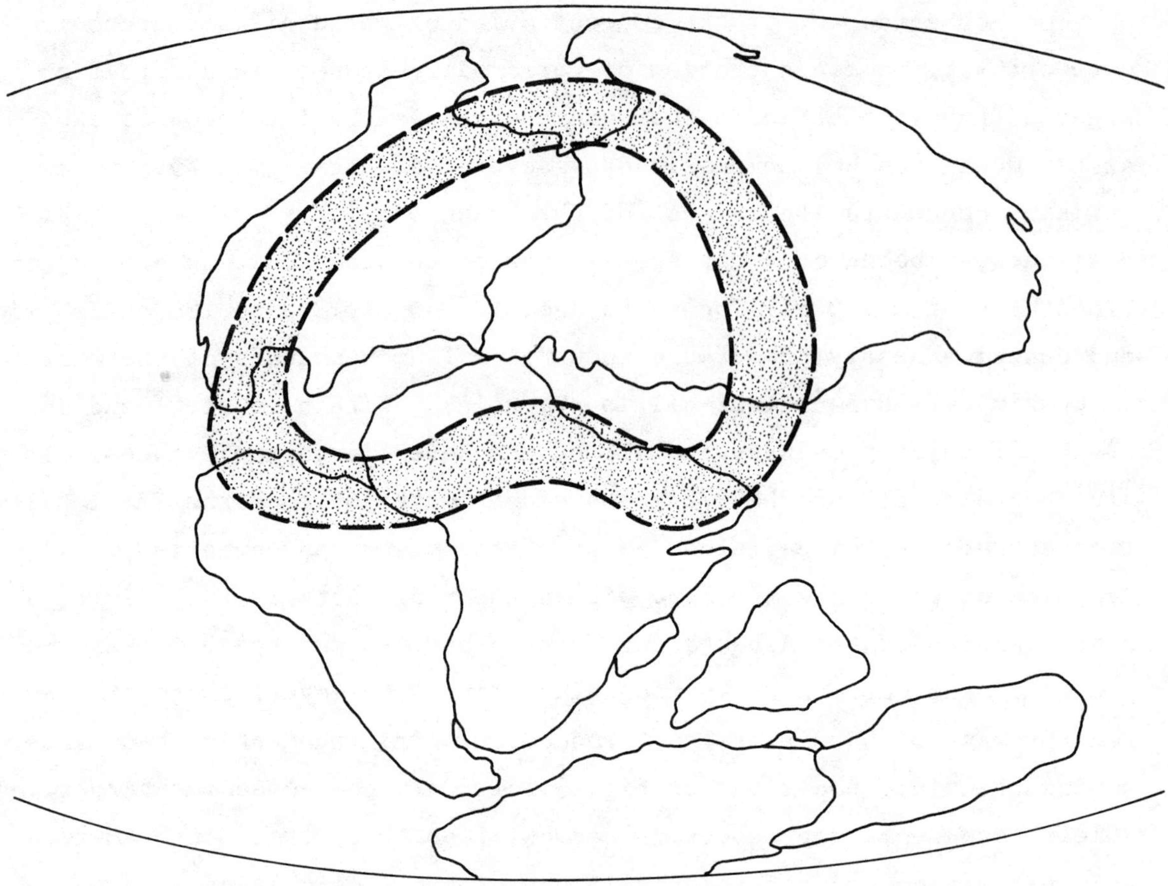

Fig. 3.1—The "ring of oil"

SOURCE: Map of the postulated position of the continental plates at the end of the Triassic Period (180 million years ago), adapted from R. S. Dietz and J. C. Holden, "The Breakup of Pangea," *Scientific American,* Vol. 223, No. 5, November 1970, p. 35.

IV. THE DISCOVERY OF GIANT OIL FIELDS

The importance of giant oil fields to world oil resources is historically unquestionable. Additions to world oil resources in the past have primarily been closely associated with the rate of discovery of giant oil fields. This section examines the rate of discovery of giant oil fields, both worldwide and in selected regions.

Figure 4.1 shows the number of giant oil fields discovered worldwide to 1975 by five-year periods (ten-year periods before 1900). The pattern is highly encouraging. Before 1925, the number of discoveries was at a low but steadily growing level. During this period, exploration was based primarily on the search for surface evidence of petroleum or on surface geological investigation, principally for surface indications of anticlines. Trend drilling and sheer wildcatting also played significant roles.*

With the development of geophysical techniques, a growing sophistication in geologic thinking, and the initiation of extensive exploration in the Middle East, the number of discoveries accelerated after 1925. This acceleration was interrupted by the Great Depression and by World War II, two periods in which exploratory drilling worldwide was substantially reduced. Since World War II, discoveries of known and probable giant fields have grown to a high plateau, averaging nine to ten per year during the past three five-year periods. (Because of the lack of information about many of the most recent discoveries, the number of known and probable giant fields discovered is the best single indicator of the rate of discovery. The apparent decline in the number of known giants discovered since the early 1960s is primarily a function of less information about many of the discoveries of the late 1960s and early 1970s.)

During the past 25 years a significant shift has occurred in the location of discoveries. The discovery of onshore and partly offshore

---

*For a comprehensive history of petroleum exploration, see E. W. Owen, *Trek of the Oil Finders: A History of Exploration for Petroleum*, American Association of Petroleum Geologists Memoir 6, Tulsa, Okla., 1975.

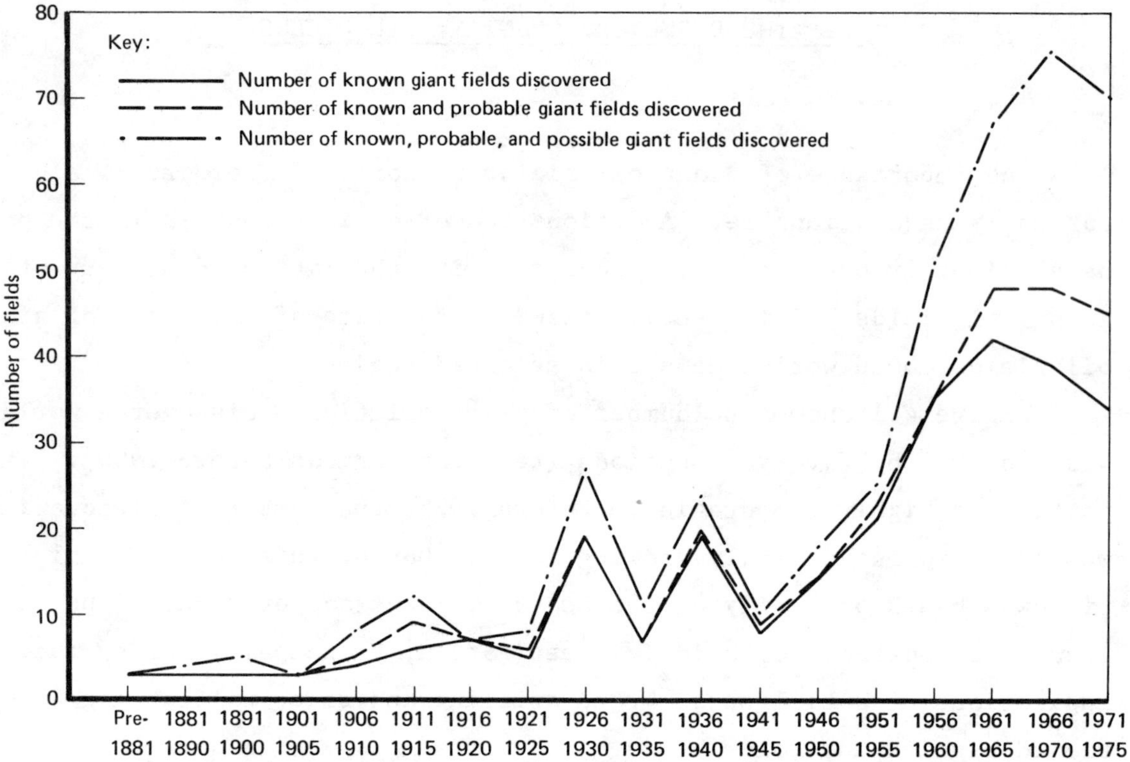

Fig. 4.1—Discoveries of giant oil fields worldwide through 1975

fields has declined sharply after reaching a peak in the early 1960s. (The number of known and probable giants discovered onshore and partly offshore in the five most recent five-year periods has been 18, 29, 36, 26, and 18, respectively.) The number of giant fields discovered offshore has steadily increased, finally surpassing onshore discoveries in the early 1970s. (The number of known and probable giants discovered offshore in the five most recent five-year periods has been 5, 6, 12, 22, and 27, respectively.)

Although the trend in the number of giant fields discovered is encouraging, the trend in the amount discovered is not. Figure 4.2 shows the amount of oil discovered in known giant fields to 1975 by five-year periods after 1900. (Total recovery in each field as of the end of 1975 is credited to the discovery year, even though all of these fields were not fully developed until subsequent years. Reliable data to credit additions to recovery to the years of development are not available.) Like the number of discoveries, the amount discovered was

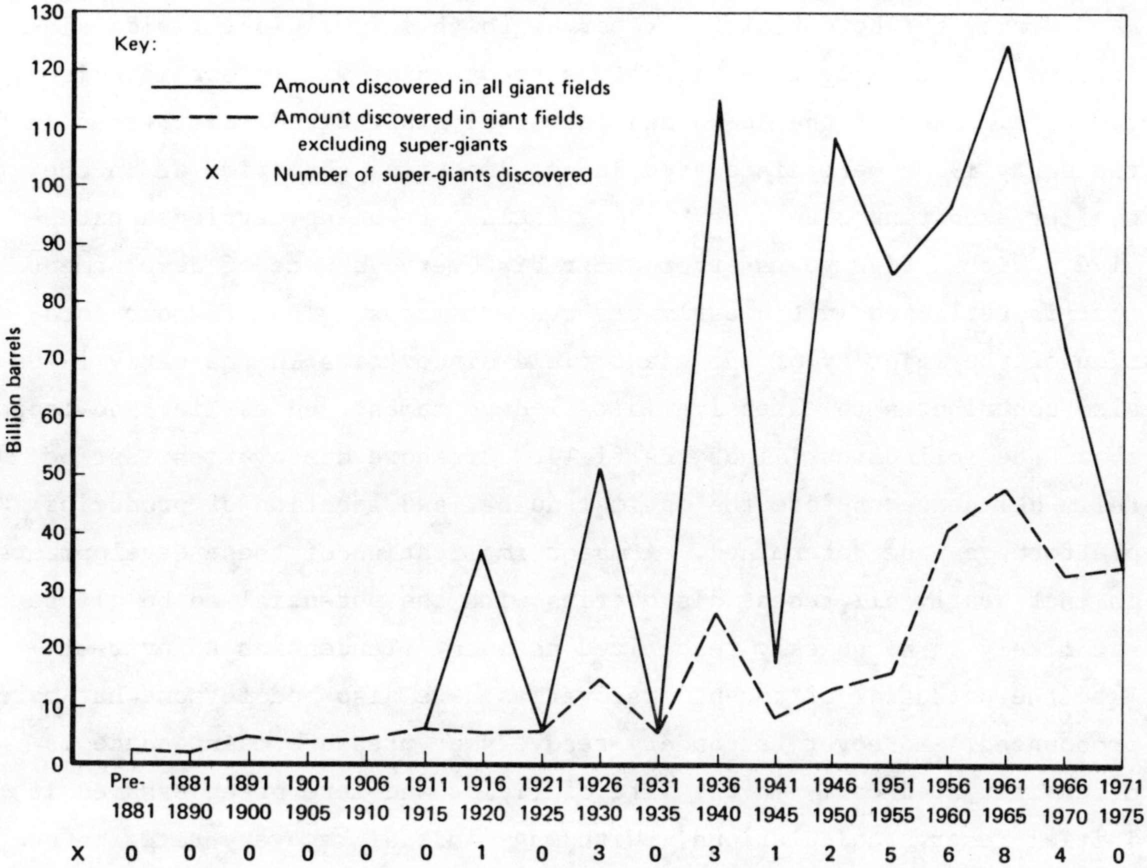

Fig. 4.2—Crude oil discoveries in known giant oil fields worldwide through 1975

at low levels until 1925 (except for the discovery of the Bolivar Coastal Field in 1917). When exploration began in earnest in the Middle East, it quickly grew to record heights, broken by the interruptions of the Great Depression and World War II. From the late 1940s to the early 1960s, the amount discovered in known giant fields was at a high plateau, averaging 20.7 billion barrels per year. Since the early 1960s, it has dropped precipitously.

The key question is whether this decline is real or is only apparent--a result of crediting the amount discovered to the discovery year. Historically, many giant fields have not been identified as such, and the full potential of fields has not been recognized in the first few years after discovery. The same phenomenon may apply here as well, making the decline only an apparent one resulting from a relative lack of information about the discoveries of the early 1970s. However, there appear to be two substantial reasons for arguing that the decline is real and not just apparent.

First, the potential for reserve growth in the giant fields discovered in the early 1970s is not as great as it was in earlier discoveries. Many of the known and potential giant fields discovered in the early 1970s were discovered in net importing countries or in the smaller exporting countries. These fields are being developed extensively within five years after their discovery, a pace of development that is reflected in the early reserve estimates. The offshore location of the majority of the giant field discoveries in the early 1970s also contributes to extensive initial development and earlier knowledge about the full potential of the field. Offshore discoveries must be fully delineated before the optimal number and location of producing platforms can be determined. A major implication of these developments is that nearly all recent discoveries with the potential to be giants are likely to be quickly recognized as such. Tendencies to overestimate the potential of recent discoveries have also become somewhat more pronounced. Moreover, secondary recovery or pressure maintenance is often initiated early in the life of fields and more often assumed in initial reserve calculations. With many initial recovery rates in new discoveries now being estimated at 40% to 50% of oil-in-place, reserve growth of more than 20% to 50% from initial estimates is highly unlikely.*

---

*The subject of reserve growth has been an area of intense controversy during the past five years. See R. G. Seidl, "Implications of Changing Oil Prices on Resource Evaluations," in Meyer (ed.), *The Future Supply of Nature-Made Petroleum and Gas*, pp. 114-117. For a highly optimistic view of reserve growth, see P. R. Odell and K. L. Rosing, *The North Sea Oil Province*, Kogan Page, London, 1975, Chap. 4. These authors use past reserve appreciation factors, derived from experience during a period of low demand and hence little incentive to develop major discoveries, especially when fields were subject to severe production restrictions (as in the United States from the 1930s to the 1960s) or lacked a market of sufficient size (the Middle East from the late 1920s into the 1950s). Such factors are methodologically inappropriate during periods in which different circumstances prevail. During the past 10 to 15 years, reserve appreciation in new fields has been substantially less. See Energy Resources Conservation Board, *Alberta's Reserves of Crude Oil, Gas, Natural Gas Liquids, and Sulphur at December 31, 1976*, Calgary, Alberta, 1977, pp. 9-3 to 9-5; and T. D. Adams and M. A. Kirkby, "Estimate of World Gas Reserves," *Proceedings, Ninth World Petroleum Congress*, Vol. 3 (Exploration and Transportation), Applied Science Publishers, London, 1975, pp. 5-6. The subject of reserve growth will be treated in greater detail in Section V.

Second, all of the major fluctuations in the amount discovered during the past 60 years have resulted from the number and size of the super-giants discovered. Discoveries of currently recognized super-giants reached a peak in the early 1960s. Since then they have dwindled to zero. The amount discovered in super-giants was at a high plateau from the late 1930s to the early 1960s (broken only by World War II) and has since declined to zero. Excluding the decline in the discovery of super-giant fields, the amount discovered in other giant fields has been relatively stable during the four most recent five-year periods. If large giant fields having a potential to become super-giants with additional investment in further recovery are included in the analysis, the pattern of discovery is not significantly altered. Fifteen known giant fields could eventually be recognized as super-giants. Only two, *Abu Ghirab-Dehluran* (Iraq-Iran) and *Samaria-Cunduacan* (Mexico), were discovered in the early 1970s.* No other known fields, recent discoveries or otherwise, appear to have a reservoir volume that indicates super-giant potential. If anything, therefore, including potential super-giants makes the decline in super-giant discoveries from the early 1960s to the early 1970s even more pronounced.

Allowing for likely reserve growth in the known giants and the increasing availability of information about potential giants, particularly in the Middle East, the ultimate amount that will be recovered from giant fields discovered during 1971-1975 is likely to be between 60 and 80 billion barrels. With the ultimate amount discovered during 1961-1965 at least 140 billion barrels (assuming the same type of adjustments), the real decline is at least 40% to 55%. The difference

---

*The other 13 (with their year of discovery and location) are *Dukhan* (1940, Qatar), *Minas* (1944, Indonesia), *Zubair* (1948, Iraq), *Khursaniyah* (1956, Saudi Arabia), *Sabriyah* (1956, Kuwait), *Amal* (1959, Libya), *Nasser* (1959, Libya), *Bibi Hakimeh* (1961, Iran), *Gialo* (1961, Libya), *Karanj* (1963, Iran), *Paris* (1964, Iran), *Rag-e Safid* (1964, Iran), and *Buzurgan* (1969, Iraq). *Khurais* (Saudi Arabia), a known super-giant discovered in 1957, could expand severalfold if it links up as speculated with *Abu Jifan, Mazalij*, and *Qirdi*. Two undeveloped potential giants, *Qayarah* (1927, Iraq) and *Salym* (1963, Soviet Union), may have the potential to become super-giants if the two respective governments are willing to invest in many low-productivity wells and sophisticated recovery operations.

is significant. The average annual amount discovered during 1961-1965 exceeds current annual world consumption by 25%. The average amount discovered during 1971-1975 is little more than half that amount.

Because the Middle East, with nearly half of the recoverable oil discovered to date, is unique among the world's petroleum provinces, the rate of discovery within it is highly significant. Figure 4.3 shows the number of giant discoveries and the amount discovered in giant fields in the Middle East by five-year periods through 1975. As a comparison of Fig. 4.3 with Fig. 4.2 indicates, most of the fluctuations in the amount discovered worldwide during the past 50 years have resulted from fluctuations in the amount discovered in the Middle East. These fluctuations in turn result from fluctuations in the discovery

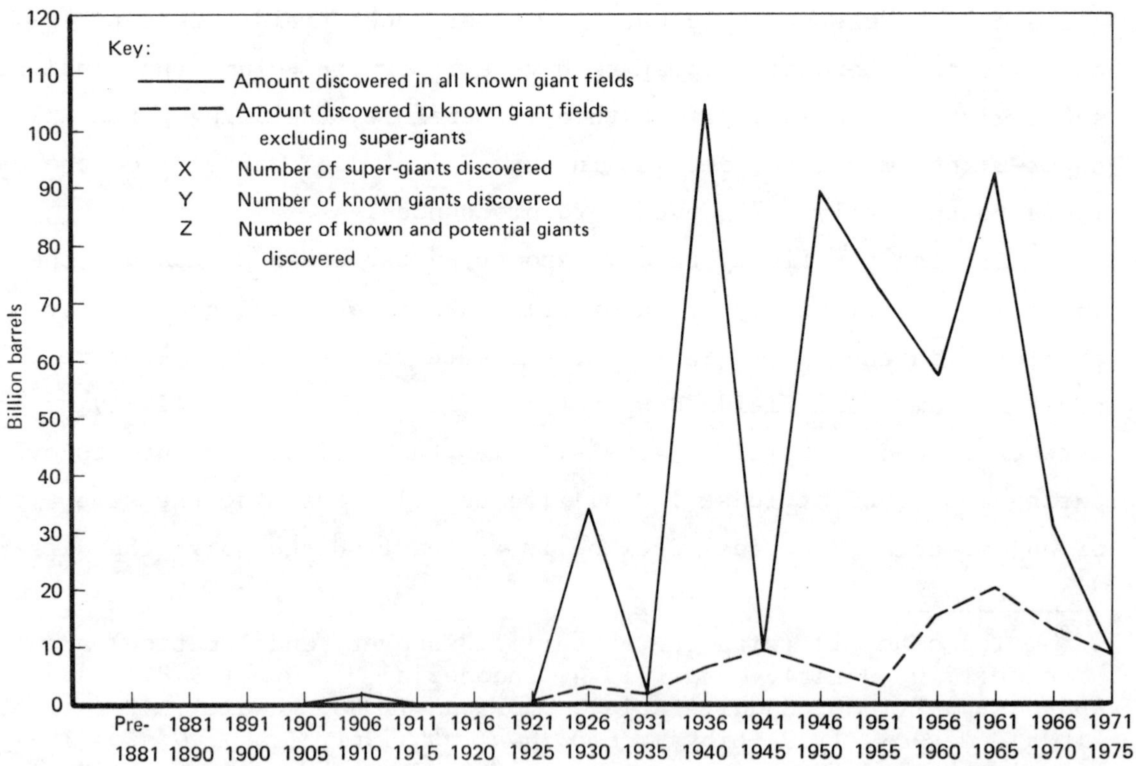

	1906 1910	1926 1930	1931 1935	1936 1940	1941 1945	1946 1950	1951 1955	1956 1960	1961 1965	1966 1970	1971 1975
X	0	2	0	3	1	1	5	4	7	2	0
Y	1	3	2	7	1	4	7	11	18	16	9
Z	1	5	2	8	1	5	7	14	29	34	26

Fig. 4.3—Crude oil discoveries in giant fields in the Middle East through 1975

of super-giant fields, which have slightly more than 80% of known recoverable Middle East resources. The amounts discovered hit their initial peaks during the periods when *Burgan* (1938) and *Ghawar* (1948), the two largest fields in the world, were discovered. A third peak was reached during the early 1960s when the number of super-giant fields discovered in the Middle East reached its height during a period of extensive exploration onshore in Iran and offshore in the Persian/Arabian Gulf. Since then, the known amount discovered in Middle East giants has dropped precipitously. To some extent, as discussed earlier, this decline is only apparent. As the potential giants discovered in the early 1970s are developed, the known amount is likely to double, triple, or even quadruple in size. But even with substantial allowances for future reserve growth, because of the disappearance of super-giant discoveries and because of the absolute decline in the number of known and potential giant field discoveries, the amount discovered in giant fields in the Middle East has declined by at least 50% and possibly by as much as 75% from the early 1960s to the early 1970s. Moreover, the decline occurred despite an increase in exploratory drilling during this same period.

The pattern of discovery outside the Middle East is more encouraging (Fig. 4.4). The basic trend in the amount discovered in giant fields other than super-giants has been one of steady growth with minor fluctuations resulting from the initiation of exploration in the Permian Basin and East Texas (late 1920s), the Great Depression, and World War II. The discovery of super-giants has clearly affected the amount discovered, but because only eight have been discovered outside the Middle East, the effect is not nearly as dramatic. Allowing for reserve growth, the amount discovered during each of the four most recent five-year periods has been at least 40 to 50 billion barrels, with perhaps a 10% to 20% real decline from the late 1960s to the early 1970s. During the same period, the number of known giant fields discovered has averaged about five per year. The number of known and probable giant fields discovered has grown steadily, totaling 24, 26, 26, and 33, respectively, during the four most recent five-year periods.

The high plateau in both the number of giant fields discovered

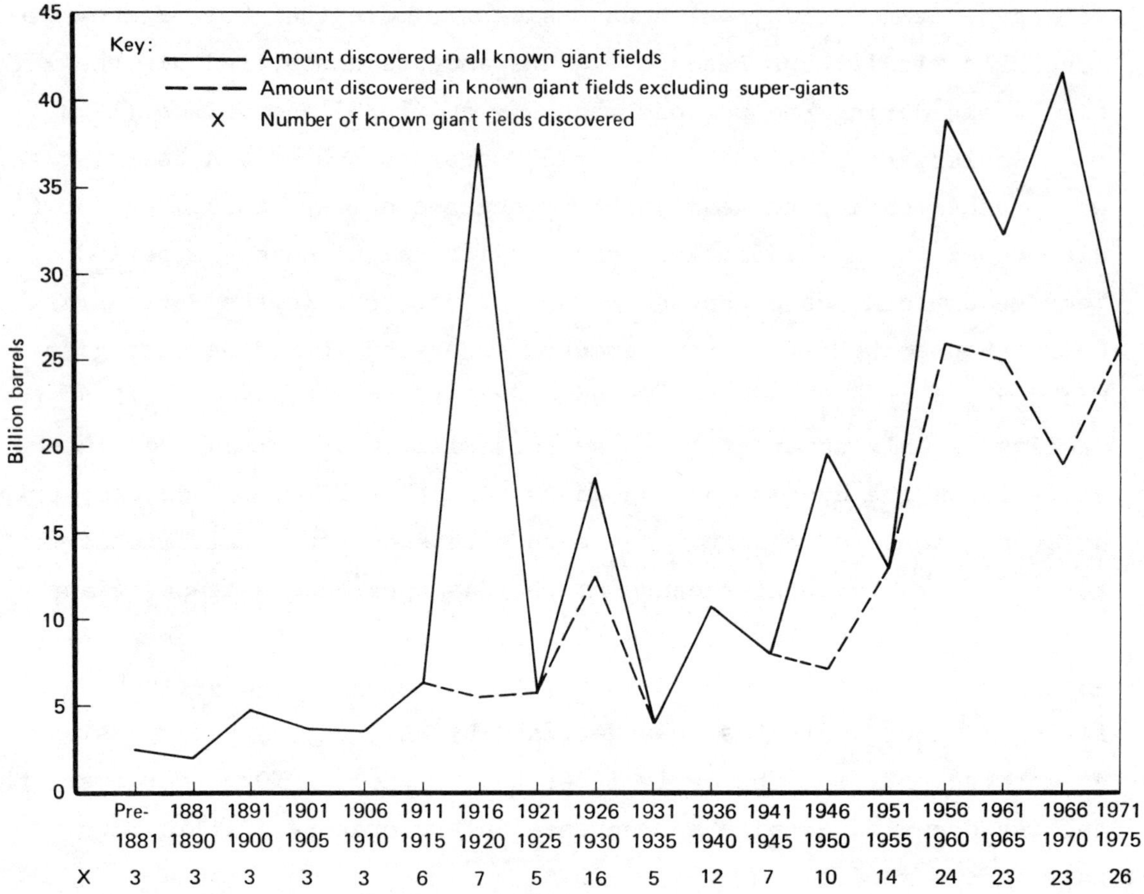

Fig. 4.4—Crude oil discoveries in giant fields outside the Middle East through 1975

and the amount discovered in them during the past 20 years has been closely associated with the discovery and development of new major provinces. During this period, the first discoveries of giant fields were made in 9 of the 21 major provinces outside the Middle East (Table 3.4): the Erg Oriental (Algeria), Niger Delta (Nigeria), Sirte Basin (Libya), and Sung-liao Basin (China) in the late 1950s; the West Siberian (Soviet Union) and North China (China) basins in the early 1960s; the North Slope (United States) and the North Sea (Norway/United Kingdom) in the late 1960s; and the Reforma area (Mexico) in the early 1970s. Of the 109 known and probable giant fields with 144+ billion barrels of identified recoverable reserves discovered outside the Middle East during the past two decades, 68 (62.4%) with 110.7+ billion barrels

(76.9%) were found in these nine provinces. The importance of new provinces for maintaining high levels of discovery is further underscored by the fact that the time required to locate most, if not all, of the major prospects in a province now takes only 5 to 10 years, assuming the absence of political constraints over the area to be explored.*

The importance of new provinces for a continued level of major discoveries is further illustrated by the discovery pattern for North America (Fig. 4.5). This continent has proved to be the most important

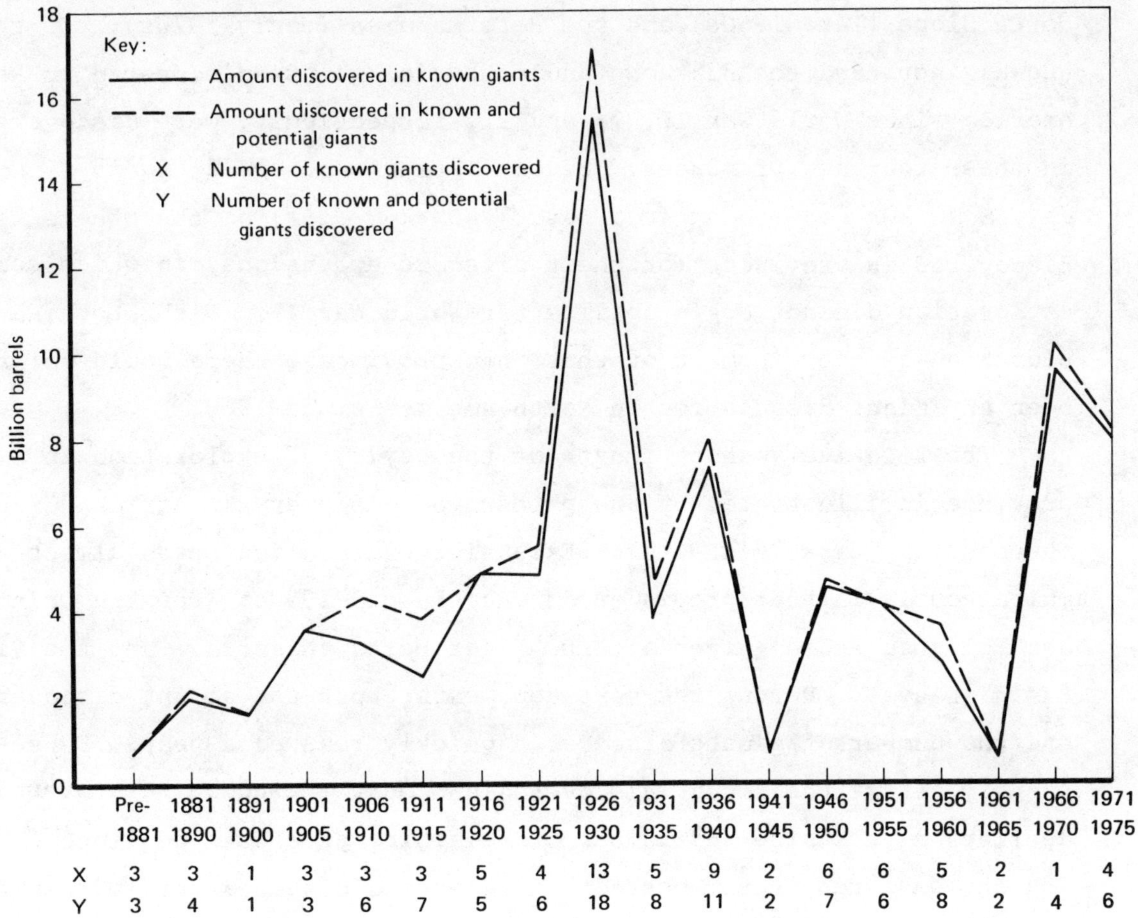

Fig. 4.5—Crude oil discoveries in giant fields in North America through 1975

---

*Klemme, "Trends in Basin Development." The subject of exploratory efficiency will be discussed in some detail in Section V.

region for petroleum exploration outside the Middle East, with 104 giant or potential giant oil fields in 24 provinces. Discoveries of giant oil fields in North America peaked during the late 1920s when major exploration began in East Texas, the Permian Basin, the onshore Mississippi Delta, and the Gulf Coast (four of the nine major provinces discovered in the continent to date). From this peak, both the number of giant fields discovered and the amount discovered fell precipitously. The reversals in this trend since World War II have resulted primarily from discoveries in the Alberta Basin and the offshore Mississippi Delta from the late 1940s to the late 1950s and the discoveries on the North Slope (late 1960s) and the Reforma area (early 1970s). Of the 24 known giants and the 33 known and potential giants discovered in North America since World War II, 18 and 21, respectively, were discovered in these four new provinces. Seven (*Altamont-Bluebell*, *Aneth*, *Hondo*, *Jay*, *McArthur River*, *San Ardo*, and *Weyburn-Midale*) of the others were discovered in provinces (or their offshore extensions) in which serious exploration did not begin until after World War II. Without the exploration and development of these new provinces, there would not have been any giant discoveries in North America since 1960.

The relatively short length of the heyday of exploration in a province is illustrated by the pattern of discovery in Africa during the past 25 years (Fig. 4.6). Extensive exploration began almost simultaneously in four provinces during the mid-1950s: the Erg Oriental and Polignac Basin (Algeria), the Niger Delta (Nigeria), and the Sirte Basin (Libya). During the next ten years, both the amount discovered and the number of giants discovered quickly reached a peak. Subsequently, first the amount discovered and next the number discovered declined as the size and number of remaining prospects declined. During the past ten years, most of the giants discovered were either fields in newly opened areas in developed provinces (*Intisar A* and *Intisar D* in the Sirte Basin and *July* and *Ramadan* in the Gulf of Suez) or fields in new provinces (*Malongo West* and *Malongo North and South* offshore in the Cabinda Embayment). The implications of these patterns of discovery for remaining discoverable world oil resources will be considered explicitly in the following section.

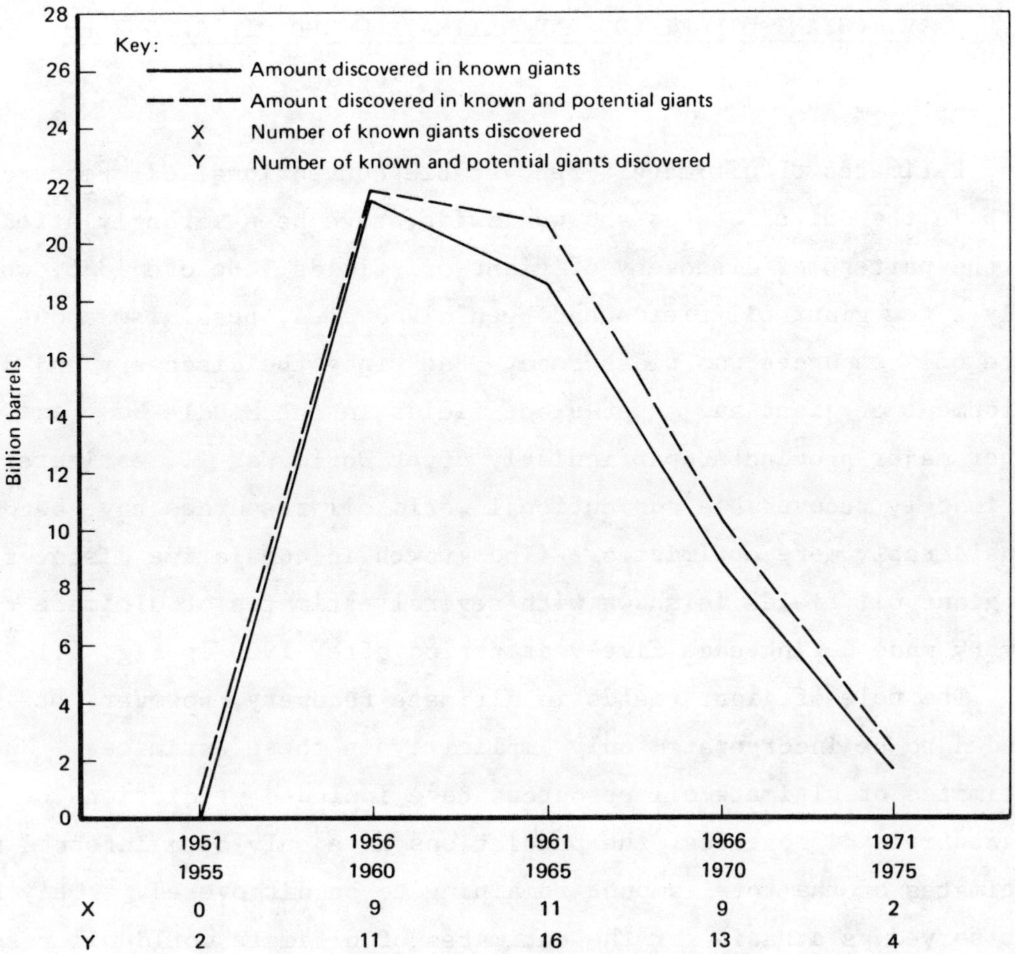

Fig. 4.6—Crude oil discoveries in giant fields in Africa through 1975

V. GIANT OIL FIELDS AND ULTIMATE WORLD OIL RESOURCES

Estimates of ultimately recoverable conventional oil resources, both in the United States and worldwide, have been strongly affected by the pattern of discovery of giant oil fields. Before 1925, when only a few giant oil fields had been discovered, pessimism about ultimate oil resources was widespread. But since the discovery and development of giant and super-giant fields in the Middle East and in other major provinces, particularly after World War II, estimates of ultimately recoverable conventional world oil resources have become considerably more optimistic. (The growth in cumulative discoveries in giant oil fields is shown with several estimates of ultimate recovery made during each five-year period after 1900 in Fig. 5.1.[*])

The role of giant fields in ultimate recovery, however, has tended to be incorporated only implicitly in these estimates. Where estimates of ultimate oil resources have included predictions of future giant field discoveries, the predictions have only been inferred from estimates of the total amount remaining to be discovered.[†] They have not served as a basis for the estimates of ultimate world oil resources. In many cases, estimates of ultimate oil resources also appear to have lagged behind the pattern of discovery and development of giant fields by up to ten years. Many of the early estimates clearly understated ultimate recovery based on discoveries made up to the time of the

---

[*] The cumulative growth in discoveries of oil in known and probable giant oil fields dated back to the year of discovery is derived from Appendix A. The estimates of ultimate recovery are taken from Moody and Esser, "An Estimate of the World's Recoverable Crude Oil Resource"; O. E. Childs, "Implications for Future Petroleum Exploration," in Meyer (ed.), *The Future Supply of Nature-Made Petroleum and Gas*, pp. 87-88; and Seidl, "Implications of Changing Oil Prices on Resource Evaluations," p. 114.

[†] For example, J. D. Moody and H. H. Emmerich estimated that 200 to 300 giant fields remained to be discovered, assuming that 75% of the 800 to 900 billion barrels that they estimated remained to be discovered would be found in giant fields. See their "Giant Oil Fields of the World," *Proceedings, Twenty-Fourth International Geologic Congress, Section 5, Mineral Fuels*, Montreal, 1972, pp. 161-167.

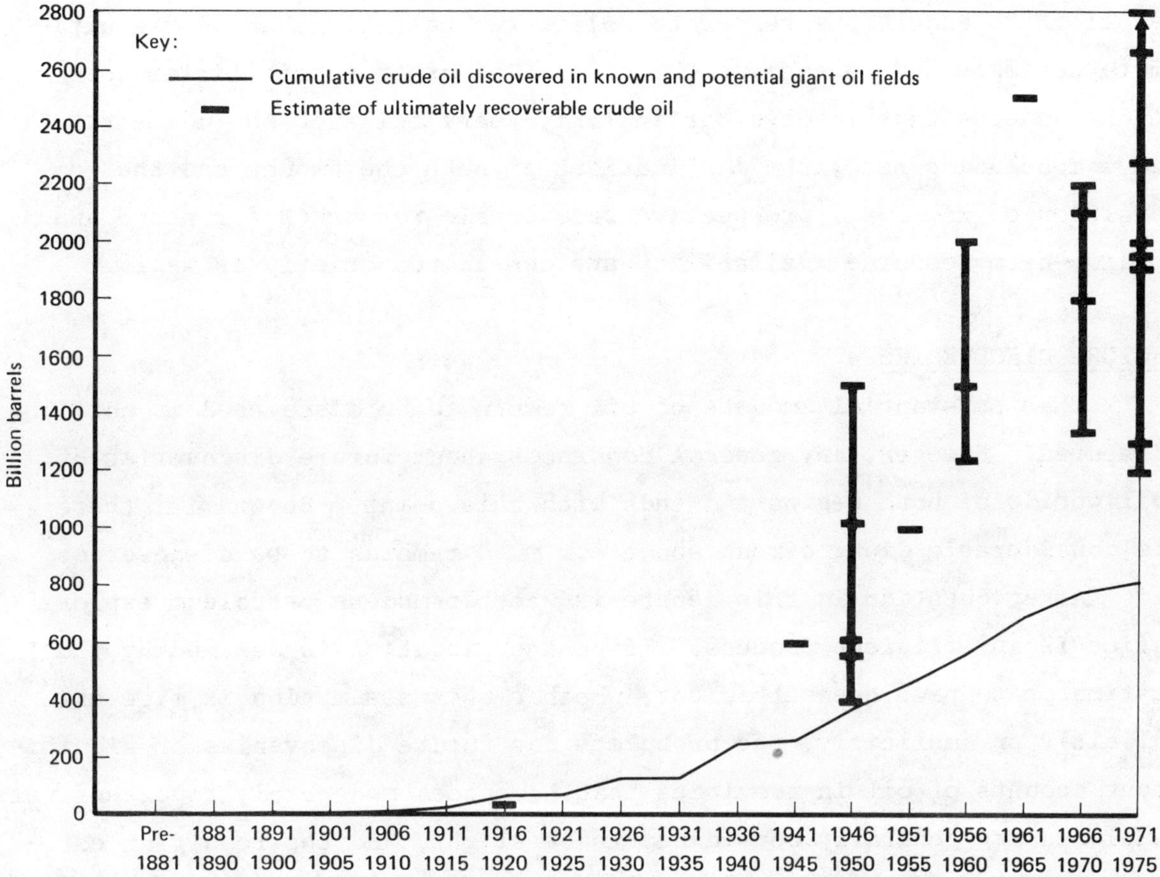

Fig. 5.1—Cumulative crude oil discovered in giant fields and estimates of ultimately recoverable crude oil

estimate. Some of the more recent estimates appear to have overstated ultimate recovery by extrapolating from the peak periods in the discovery of super-giant oil fields and in the development of the Middle East.

This section provides an explicit evaluation of ultimately recoverable conventional crude oil resources based primarily on an analysis of the potential contributions of giant oil fields. Two main topics are addressed: future discoveries and reserve growth. The estimates of potential additions from each of these two sources are then summarized in an overall assessment of ultimately recoverable conventional crude oil resources. Because the overall assessment is based on the dissaggregated analysis, each of the several aspects of the two general topics is discussed in some detail. The discussions are deliberately explicit to enable the reader to follow the reasoning behind the ultimate estimate and to examine the effects of other possibilities for those aspects that involve particularly heavy reliance on judgment. As a concluding note, the implications of both the amount and the composition of known and prospective recoverable resources for near- and medium-term resource availability are considered briefly as well.

FUTURE DISCOVERIES

That substantial amounts of oil remain to be discovered is not disputed. However, any general consensus about future discoveries of oil worldwide both begins and ends with this point. Beyond it, there is considerable disagreement about *how much* remains to be discovered.*

A key question in this debate is whether modern petroleum exploration is an efficient process. If past exploration is assumed by the estimator to have been efficient, whether this assumption is made explicitly or implicitly, the prospects for future discoveries of significant amounts of oil in provinces that have been moderately to extensively explored are generally considered to be slight, and the resulting estimates of ultimate recoverable resources are likely to be pessimistic as well. If exploration is assumed to have been inefficient, significant future discoveries within most of the known producing provinces are still a real possibility and estimates of future discoveries can be optimistic.

The available empirical evidence suggests that modern petroleum exploration is generally efficient. More specifically, exploration for giant fields appears to be a highly efficient process. Exploration for smaller fields appears to be moderately efficient. Operationally, high efficiency in exploration will be defined here to mean that, except for the handful of provinces that are very large in area such as the Arabian-Iranian and West Siberian basins, or for the few deltaic

---

*The range of disagreement is illustrated in Fig. 5.1, and in the estimates provided by the Conservation Commission, World Energy Conference, *Report on Oil Resources: 1985-2020, Executive Summary*, Istanbul, 1977. The estimates published in the latter source range from 1200 to 3500 billion barrels ultimate.

provinces with many large fields, a well-designed exploratory program is likely to result in the discovery of 90% to 100% of the giant fields in any given province with the drilling of no more than 25 to 200 new-field exploratory wells, the exact number depending primarily on the area and geologic characteristics of each particular province. When exploratory drilling occurs at reasonable rates, which it has in every province except the Middle East, high efficiency means that nearly all of the giant fields will be discovered in the first five to ten years of exploratory drilling* unless there are limitations on the area that can be explored. High efficiency also implies that if the first 10 to 30 new-field exploratory wells are dry holes, the province has a very low probability of ever becoming a significant source of oil, assuming that the location of the wells was determined by an adequate geological and geophysical survey and that the wells were drilled to sufficient depths.

There are several reasons why modern petroleum exploration is able to locate giant fields efficiently:

1. Nearly all giant fields are found in large, detectable traps.† They therefore provide obvious targets for exploratory drilling.
2. Improvements in geologic knowledge and exploratory technology are enabling the petroleum industry to locate the large, detectable traps with increasing effectiveness before drilling.
3. Contemporary exploratory philosophy concentrates on the search for giant fields in order to recoup the large front-end costs of exploring and developing a new province as soon as possible.‡

This threefold combination means that the major uncertainty in modern petroleum exploration is whether the traps that are located before

---

*Klemme, "Trends in Basin Development."

†Detectable traps are defined as traps whose location could have been determined by surface or seismic mapping. Klemme, "Structure-Related Traps Expected To Dominate World-Reserve Statistics," found that 94.4% of the giant fields he surveyed were detectable traps.

‡Ivanhoe, "Evaluating Prospective Basins."

drilling will contain commercial quantities of oil and gas. As recent U.S. experience in the eastern Gulf of Mexico and the Gulf of Alaska has indicated, large structures can produce little more than disappointing results.

A few examples from some of the larger provinces illustrate the efficiency of modern petroleum exploration. Exploratory efforts in the Permian Basin, even though they occurred when modern technology was still in its primitive stages, found 18 of its 23 known and potential giants during the first 15 years of exploration (13 during the first 8 years). Exploration in the Middle East has been amazingly efficient. The first wells drilled in Bahrein and Qatar discovered *Awali* and *Dukhan*, respectively. The second well drilled in Kuwait discovered *Burgan*. The first wells drilled offshore Abu Dhabi and Dubai discovered *Umm Shaif* and *Fateh*, respectively. After the discovery of *Dammam* in 1938, the 55 new-field exploratory wells drilled in Saudi Arabia from 1939 to 1976 resulted in at least 37 discoveries (7 of the first 15 new-field exploratory wells drilled after the *Dammam* discovery found super-giants). Of the 74 new-field exploratory wells drilled by the Consortium/OSCO in Iran between 1958 and 1976, 36 were discoveries, 9 were suspended (as either gas discoveries or marginally commercial oil discoveries), and 29 were abandoned (with at least 8 of the latter finding oil or gas that would be producible under anything other than Middle East oil economics). The first four new-field exploratory wells drilled in southern Iraq were discoveries, two being super-giants. The 16 known and potential giant fields of the Sirte Basin in Libya were all found between 1959 and 1968, 10 being discovered from 1959 to 1961. Despite indications that Soviet exploratory technology is inferior to Western technology, the five largest oil fields in Western Siberia were found in only six years (1961-1966). Exploration in the North Sea and the Reforma area of Mexico appears to be highly efficient as well.*

---

*The principal counterargument to the hypothesis that modern petroleum exploration is efficient is that a random search process can approximate historical drilling results. H. W. Menard and G. Sharman ("Scientific Uses of Random Drilling Models," *Science*, Vol. 190, No.

In evaluating the potential for future discoveries, this section considers several topics: the discovery of new provinces, the discovery of super-giants, the discovery of other giant fields, and discoveries of non-giant fields. In each case, the relevance of exploratory efficiency for the analysis of future potential is incorporated into the discussion. As part of the evaluation, quantitative predictions of future discoveries are provided. These predictions are the author's best judgments, within the unavoidable range of uncertainty, based on analogies from observed trends and patterns. The rationale for the predictions is made explicit so that readers can make their own judgments as to whether the predictions are reasonable and defensible.

Province Discovery and Development

The past discovery of crude oil has depended primarily on the discovery and development of a few major provinces. Twenty-two provinces

---

4212, October 24, 1975, pp. 337-343) found that the historical record of exploration in the United States corresponds in general form for long periods to the results of a random search. Their results do not, however, constitute a counterargument to the hypothesis that *modern* petroleum exploration *worldwide* is highly efficient in the search for giant fields, primarily because the history of petroleum exploration in the United States is an atypical example. Exploration in this country began in what have proved to be the less productive provinces. The model used by Menard and Sharman assumes that the choice of area is random. Unlike the rest of the world, scientific methods of exploration were not used in the United States until 1910 to 1920, and most of the major provinces in this country were opened up when exploratory knowledge and technology were still in their primitive stages. The widespread use of surface geology and the introduction of geophysics correlate in time with the high rate of discovery of giant fields between 1920 and 1935 in the Los Angeles, Permian, East Texas, Texas Gulf Coast, and Mississippi Delta provinces. This period is the only one in which the historical discovery rate was greater than the various random discovery rates generated by the Menard-Sharman model. Most early exploratory wells were quite shallow. The Menard-Sharman model assumes an average depth of 5000 feet per well. Most of the exploratory drilling in the United States has been performed by independent oil companies. Menard and Sharman note that the major oil companies, who use scientific methods more extensively, are considerably more successful. Most petroleum exploration outside the United States has been conducted by majors or state companies using modern technology. Because of differences in historical and institutional circumstances, petroleum exploration abroad is proving to be markedly more efficient than it has been within the United States.

that have or are likely to have an ultimate recovery of 10 billion barrels or more account for 85% of the oil discovered and made recoverable to the end of 1975. Another 13 that have an ultimate recovery of 3 to 10 billion barrels each account for 5% of the oil discovered and made recoverable to the end of 1975. The remaining 90 producing provinces account for the other 10%. Because the time to discover most of the recoverable resources in a province generally has taken only 10 to 20 years and because this period appears to be shortening to only 5 to 10 years,* very large new discoveries, if they are to occur, will depend primarily on the discovery of new major provinces.

Whether any major new provinces remain to be discovered is intrinsically a highly speculative matter. The past record does, however, provide a few clues as to the probability of one or more being discovered. First, if a major new province is to be discovered, it is highly probable that it will be a province in which little or no exploratory drilling has occurred to date. Since modern exploratory and drilling technology has become available, the existence of major reserves in the known major provinces has become apparent within a few years after the initiation of exploratory drilling. The search for major new provinces can thus be concentrated on the 125 or so known provinces that have yet to be moderately or extensively explored. The only possible exception among the moderately to extensively explored provinces is the Gulf of Suez.

Second, of the 8 province types--interior, cratonic intracontinental-composite, graben or rift, extracontinental, stable coastal or pull-apart, intermontane, delta, and deep oceanic†--some are more likely candidates for major provinces than others. Of the 7 types that have been explored (excluding deep oceanic basins), major provinces have been found in 5. None of the 26 interior basins nor the 53 stable coastal or pull-apart basins that have been explored has proved to be a major oil province. Only one, the interior Illinois Basin, has

---

*Klemme, "Trends in Basin Development."

†For a detailed definition, description, and analysis of basin types, see Klemme, "The Giants and the Supergiants" and "World Oil and Gas Reserves."

an ultimate recovery of 3 to 10 billion barrels. Three more producing basins--Williston and Central Kansas (both interior) and the Congo (coastal pull-apart)--may also become that large. Only 5 of the 79 even have known or potential giant oil fields. However, 17 of the 57 moderately to extensively explored cratonic intracontinental-composite (30%), 7 of the 28 graben or rift (25%), and 12 of the 34 extracontinental (35%) provinces either have or are likely to have at least 3 billion barrels of ultimate recovery. Of the three types, 9%, 14%, and 21%, respectively, of the explored provinces are major provinces. More important, nearly all of the largest in area of each of these three types are highly productive. Of the 13 partially to extensively explored deltas, 2 are major provinces (15%) and another is likely to have 3 to 10 billion barrels of ultimate recovery. However, because the source material for the generation of hydrocarbons in deltas is continental in origin, most of their recoverable hydrocarbons are likely to be gas rather than oil. Although 23% of the partially to extensively explored intermontane basins have known or potential giant oil fields (17 of 74), only 4 extremely rich ones are major provinces and only 2 more have an ultimate recovery of 3 to 10 billion barrels. Generally, the small area of the intermontane provinces precludes their being major provinces.

Extrapolating from past history, the most probable candidates for future major provinces are the remaining relatively unexplored cratonic intracontinental-composite, extracontinental, and graben or rift provinces with a large surface area. The relatively unexplored intracontinental-composite and extracontinental provinces with a large area are limited almost entirely to the larger provinces within or bordering the Arctic Ocean: Vilyuy-Lena, Lena-Anabar, Yenisei-Khatanga, North Yakutia, East Siberian Sea, Laptev Sea, Chukchi Sea, Beaufort Shelf, Sverdrup, and Barents Sea. Because of the severe Arctic environment, these provinces will be very expensive to explore and develop. Because of the high cost, only highly productive giant fields will be economic to develop, assuming that the technology to do so safely can be achieved for the offshore Arctic areas with a year-round pack ice cover. The only other possible candidate for a future major province of these two types is the offshore portion of the Malvinas (Falkland) Basin. Among the remaining relatively unexplored graben or rift basins, the only ones with a large surface area are the Gulf of Thailand and the rift basins of East Africa. The few exploratory wells drilled already in the Sverdrup Basin and the Gulf of Thailand have discovered several major gas fields, suggesting that these two possibilities may only prove to be important natural gas provinces.

The other province types also offer a few major province possibilities. Nearly all of the major interior basins have been at least moderately explored. Irkutsk in Central Siberia appears to be the only lightly explored province of this type with the area and sedimentary volume to become a major province, but drilling results thus far suggest that it will be a natural gas province, not an oil province. Based on the historical record alone, the probability of a stable coastal or pull-apart province becoming a major province is practically nil. However, an extensive exploration effort is under way in many of these provinces (most of which have only been lightly explored to date), and a few positive surprises could emerge by 1985. Some of the better possibilities here appear to be the Red Sea, the West Shetland Basin, the Norwegian Coastal Basin, the Labrador Shelf, the Kiangsu, and the Liu-Chou provinces. Of the unexplored intermontane provinces that could prove to be significant, the best prospects appear to be the few provinces with thick sediments in the Bering Sea and the Sea of Okhotsk. One or two large deltas may also prove significant. The knowledge available for predicting whether any of the deep oceanic basins will be major provinces is too slight to support a prediction, given the near absence of drilling efforts to date. Economic considerations will, however, be a major constraint on their potential.*

A third factor to consider in evaluating the prospects for significant new provinces is that the rate of opening up new provinces appears to be declining (dating the opening up of a province from the discovery year of its first known or potential giant field). Since 1970, only one major province (Reforma) has been opened up (compared with eight between 1956 and 1970). Only one other province (Cambay) with a potential of

---

*For a similar analysis of possibilities by basin type, see Klemme, "Giant Oil Fields Related to Their Geologic Setting."

3 to 10 billion barrels was opened up as well (also compared with eight between 1956 and 1970). This decline is attributable to the fact that nearly all of the accessible, inexpensive provinces have been at least moderately explored. Most of the more attractive remaining possibilities will be very expensive to explore and develop. Some will require new technology if they are to be developed, for example, the Arctic Ocean provinces with water depths greater than 50 to 100 feet and the other offshore provinces with water depths greater than 1000 feet. The combination of cost and technological constraints will mean a slow rate of development if any of these provinces prove to be significant.

Because of the number and characteristics of the remaining possibilities and assuming the historic probabilities that an explored province will become a major province, the likelihood that a relatively unexplored province of a given type will become a major province is small. This assessment is not only the result of the usual constraints of adequate source material for hydrocarbon generation, preservation of this material, sufficient temperature to convert it to petroleum, the presence of porous and permeable traps with adequate seals to receive the generated petroleum, and the preservation of these traps through geologic time, but also of the new constraint of the high cost of exploration and development in most of the remaining prospects. The simple probabilities by type of province only suggest an expected value of six more major oil provinces. A comparison of the surface area, sedimentary volumes, and likely hydrocarbon types (i.e., crude oil or natural gas) of the most likely prospects with the known major provinces suggests an even lower number. In the author's judgment, these considerations suggest that the most likely estimate of new major provinces that remain to be discovered is only two to four. The number will be larger only if some of the stable coastal or pull-apart provinces prove to be major provinces.

The Role of Super-Giants

The growth in known recoverable oil resources and in annual world production during the past half-century has resulted primarily from the discovery and development of super-giant oil fields. These fields provided the bulk of the easy and extremely inexpensive additions to reserves and production capacity that produced the real decline in world oil prices in the 25 years before 1973. They account for the preeminence of the Middle East among all the producing regions of the world. The few super-giants discovered outside the Middle East have each played a major role in the history of the oil industry in each of their respective countries.

The importance of super-giant fields is paralleled by their rarity. Only 33 were identified in this survey, 25 of which are in the Middle East. Another 15, 10 of which are in the Middle East, were identified as potential super-giants. The 33 known super-giants were estimated to contain 512.6 of the 1011.5 billion barrels of known recoverable oil. No province other than the Middle East contains more than one (although the Sirte Basin may end up with several super-giants if enhanced recovery can be applied successfully in several of its largest fields).

Super-giant oil fields and major provinces are closely associated. Every super-giant field or potential super-giant field is in a major province. Nearly every major province has a super-giant field, a potential super-giant field, a nonconventional deposit of super-giant dimensions, or a combination of adjacent fields adding up to super-giant dimensions. The Arabian-Iranian, Sirte, Maracaibo, West Siberian, Volga-Ural, North Slope, Erg Oriental, Sung-liao, and East Texas provinces contain super-giant fields. The Reforma and Central Sumatra provinces have potential super-giant fields. The Alberta and Maturin provinces have oil (tar) sand deposits of super-giant dimensions. Five other major provinces--the San Joaquin, Los Angeles, Permian, North Sea, and North Caucasus-Kopet Dag--have groups of adjacent giant and large fields that add up to super-giant size.* The Mississippi and Niger

---

*These groups are the *Midway-Sunset, Buena Vista, Elk Hills, McKittrick,* and *South Belridge* complex (San Joaquin), the *Wilmington Trend, Huntington Beach, Long Beach,* and *Seal Beach* complex (Los Angeles), the *Goldsmith-Andector, McElroy-Dune, TXL, North Cowden, South Cowden, Fullerton, Sand Hills,* and *Jordan-Waddell-Penwell-Edwards* fields on the east side of the Central Basin Platform (Permian), the *Statfjord, Brent, Ninian, Thistle,* and *Dunlin* group (North Sea), and the groups of oil fields on the Apsheron and Mangyshlak peninsulas (North Caucasus-Kopet Dag).

deltas with their many large fields, the Gulf Coast province with its numerous salt dome structures, and the heavily faulted North China province are the only major provinces that do not fit into one of these groups.

The role of super-giants in future discoveries does not appear to be promising. Since the peak of the early 1960s when eight known super-giants and five potential super-giants were discovered, the discovery rate has plummeted to only two potential super-giants in the early 1970s. This sharp decline is closely associated with the parallel decline in the opening up of new major provinces. Because super-giant fields are very large in area, the smallest area of any of those known being 100 square miles, and in general are found on prominent regional features within a province, they provide very obvious targets for exploration when drilling begins. Every super-giant field discovered outside the Middle East has thus been discovered relatively early in the exploration of its province. When the timing of exploration is measured by the number of exploratory holes drilled, the same generalization holds for the Middle East as well.

A few possibilities still exist for the discovery of super-giants in known producing provinces. Within the Middle East there are still some promising areas that have not been drilled sufficiently, most notably the retained areas in the Rub al Khali of Saudi Arabia and much of central and southern Iraq. Outside the Middle East, the possibility of finding super-giants in any province that has been at least moderately explored is slim. The best chance appears to be in the Reforma area of Mexico in a combination of tangential structures with very thick reservoirs. There are lesser possibilities in Western Siberia (although the Soviet emphasis on drilling the largest structures first makes this a slight chance) and in the Gulf of Venezuela (the lightly drilled northward extension of the Maracaibo province). None of the explored non-major provinces appears to offer any prospects for future super-giant discoveries. Considering the remaining possibilities in the currently producing provinces, there is a reasonable probability of two to six more super-giant discoveries in the Middle East and one to two elsewhere, for a total of three to eight future super-giants in these provinces.

Predictions as to whether any super-giants will be discovered in the nonexplored or lightly explored provinces are subject to the usual surprises, both positive and negative, that have characterized the history of oil exploration. If a super-giant is discovered in a new province, this province is highly likely to be a major province. Assuming that the previous judgment of only two to four major new provinces is correct and assuming the historic probability that roughly 50% of major provinces will have a known or potential giant oil field, only one to two super-giant fields are likely to be found in new provinces.

The estimated number of super-giant oil fields remaining to be discovered in both producing and new provinces ranges from four to ten. The average size of these super-giants will probably be between 7.5 and 10.0 billion barrels. This is smaller than the average size of known super-giants (15.5 billion barrels) both because future Middle East super-giant discoveries are likely to be smaller than the known Middle East super-giants and because super-giants outside the Middle East, which constitute a larger percentage of projected discoveries than of known super-giants, are smaller on average than the known Middle East super-giants. The total amount predicted to be forthcoming from future super-giant discoveries thus ranges from 30 to 100 billion barrels.

The Role of Other Giant Fields

Giant fields other than the super-giants contain 25% to 30% of the known recoverable crude oil in the world. The 239 known giants other than super-giants contain 263.5 of the 1011.5 billion barrels of identified recoverable crude oil resources. Potential giant fields contain another 43.1 billion barrels.

For the three most recent five-year periods, the discovery rate of giant oil fields other than super-giants has been at its historical peak. During the early 1960s, the late 1960s, and the early 1970s, 34, 35, and 34 known giant fields other than super-giants were discovered in each respective period. The number of known and probable giant discoveries other than super-giants was 40, 45, and 45. The number of known, probable, and possible giant discoveries was 59, 71, and 70.

There are several reasons for predicting that this discovery rate will begin to decline either in the late 1970s or the early 1980s. Several aspects of the discovery trend provide the basis for this prediction. Onshore discoveries have already begun to decline sharply. The number of known and probable giant fields other than super-giants discovered onshore peaked in the early 1960s at 31. By the early 1970s, it had dropped back to 18. Offshore giant discoveries probably peaked at 27 in the early 1970s and are likely to begin to decline either in the late 1970s or the early 1980s.

These trends are closely associated with the discovery and development of major provinces. Of the known and probable giant discoveries other than super-giants, 35 of 40 in the early 1960s, 35 of 44 in the late 1960s, and 38 of 45 in the early 1970s occurred in seven, seven, and six major provinces. Another 4, 8, and 6 occurred in provinces that have or are likely to have an ultimate recovery of 3 to 10 billion barrels. The decline in onshore discoveries is closely associated with the gradual completion of the primary discovery process in several of the major onshore provinces (Erg Oriental, Sirte, Niger Delta, and Western Siberia). By the mid-1970s, the Arabian-Iranian Basin and the Reforma area were the only major onshore provinces continuing to record multiple giant field discoveries. Offshore discoveries peaked in the early 1970s because of the high rate of giant field discoveries in the North Sea. These more than offset the greater than 50% decline in the number of offshore discoveries in the Arabian-Iranian Basin from the late 1960s to the early 1970s.

Two of the known major provinces still have an excellent probability of continued discoveries of giant fields. Because of the relatively low number of exploratory wells drilled and the continued high success rates of drilling, the Arabian-Iranian Basin is likely to provide many more giant discoveries, although they will be considerably smaller than those of the past when most of the super-giants were discovered. Much of Iraq, the more mountainous areas of southwest Iran, and eastern Saudi Arabia offer the best prospects. A few offshore opportunities remain as well. The Reforma area, with many identified structures still undrilled, offers possibilities for numerous giant field discoveries both onshore and offshore. Although most of these structures are relatively small in area, their great thicknesses make them obvious giant candidates.

The prospects for future giant discoveries in the other major provinces are mixed. The North Slope (onshore and near offshore), Maracaibo (especially the Gulf of Venezuela), North Sea, West Siberian, North Caucasus-Kopet Dag (the offshore Caspian Sea area), Sung-liao, and North China provinces have a good chance of a few more giant discoveries each. The Los Angeles (offshore), Mississippi Delta (deep-water offshore), Maturin (offshore), Sirte, Niger Delta, and Central Sumatra provinces have a modest probability of one or two more giant field discoveries. The Alberta, Volga-Ural, and Erg Oriental provinces may fit into this category as well, but they appear to be more comparable to the San Joaquin, Permian, East Texas, and Gulf Coast provinces, in which the probability of future giant oil field discoveries is practically zero. Among the known producing provinces that have or are likely to have an ultimate recovery of 3 to 10 billion barrels, the Gulf of Suez offers good prospects of future giant field discoveries. The Oriente, Pechora, Polignac, Dzungurian, Szechwan, Tsaidam, Cambay, and Mahakam provinces also have reasonable possibilities of future giant discoveries. If any new major or important provinces are discovered, each is certain to have several giant fields. A few single giant fields will probably be discovered in other provinces.

The likely decline in the rate of giant field discoveries, derived from an analysis of the composition of discovery rates between 1961 and 1975 and from a consideration of the remaining possibilities for future discoveries, provides a basis for predicting future discoveries of giant fields other than super-giants. Because several decline rates are possible, there will probably be 125 to 175 future (post-1975) giant field discoveries. Any number less than 100 to 110 implies a very rapid decline in the future discovery rate, a possibility that is remote given the number of remaining prospects. Any number more than 200 implies either (1) a very high number of future discoveries in the known major provinces, a possibility that is unlikely because it requires an unprecedented reversal in historic exploratory trends or (2) the discovery of more Reforma-type provinces with a large number of unusually thick structures that are small in area. Because Reforma thus far is unique in this respect among all the explored provinces, this contingency is improbable but not impossible.

Currently, the average size of the 239 known giants other than super-giants is 1100 million barrels. Because the majority of future discoveries of giant fields is likely to be offshore and known offshore giant discoveries have been smaller on average than known onshore discoveries (975 as compared with 1135 million barrels), and because future giant field discoveries in known provinces with giant fields will probably be smaller on average than past giant field discoveries, the average size of future giant field discoveries other than super-giants is estimated to be between 800 and 1000 million barrels. The total amount in giant fields other than super-giants that remains to be discovered is thus estimated to be 100 to 175 billion barrels.

The Role of Non-Giant Fields

Non-giant oil fields currently contain only a small proportion of known recoverable crude oil resources. Of the estimated 1011.5 billion barrels discovered and made recoverable to date, roughly 90 to 95 are in an estimated 500 to 525 large non-giant fields (100 to 500 million barrels each) and another 97 to 102 are in medium, small, and very small non-giant fields (see Table 3.2).*

In general, the large non-giant fields share several important characteristics with giant fields. They are relatively small in number. Most are among the larger fields of the province in which they are found. Most are found early in the history of exploration of a province, although the average period of discovery is longer than that for giants because of the smaller size and typically less attractive economics of the large non-giants. Large fields are also concentrated in the major

---

*Medium-size non-giant fields will be defined here as having 25 to 100 million barrels; small fields will be defined as having 10 to 25 million barrels; and very small fields will be defined as having less than 10 million barrels.

provinces, although to a lesser degree than the giants. (The major provinces have 83.5% of the known giants, 73.2% of the potential giants, and approximately 65% of the large non-giant oil fields.) The distribution of large non-giant fields among the major provinces differs from the distribution of giants. Instead of being concentrated in the largest major provinces, large non-giant fields are most prominent in the two major delta provinces. Because of their more even field size distribution, the Mississippi and Niger deltas each have roughly 10% of all large non-giant fields. Most of the other major provinces have between 10 and 40 large and potentially large non-giant fields, the exceptions being the few provinces (East Texas, Maracaibo, Erg Oriental, and possibly the North Slope) where production and reserves are concentrated in a single super-giant field. Comprehensive information about large non-giant fields in the major provinces of the Soviet Union and China is not available.

As a first step in predicting future large non-giant discoveries, one could assume that because of the similarity in characteristics between giant fields and large non-giant fields and because discoveries of large fields lag slightly behind those of giant fields, the proportional increase in the number of large non-giant discoveries will be somewhat greater than the probable increase in the number of known giants. This approach would suggest that 300 to 450 large non-giant fields will be discovered in the future. (This calculation assumes (1) that 350 giants other than super-giants have already been discovered in known, probable, potential, and unidentified giant fields; (2) that 125 to 175 more such giants will be discovered, for an increase of 36% to 50%; (3) that 600 large non-giants have already been discovered, including both some potential giants that will prove to be only large fields and some potential large fields; and (4) that the total number of non-giants will increase by 50% to 75%.) This estimate of future large non-giant discoveries is high enough to include both a sizeable number of additional discoveries in the known provinces and discoveries of several new major provinces with significant numbers of large non-giant fields.

Two factors, however, suggest that future discoveries could be greater than the range suggested by this approach. First, and most important, a substantial number of large non-giant fields in subtle traps (i.e., traps that are non-detectable by seismic surveying) could be discovered. In the early years of exploration of a province, most of these traps are likely to be missed because they do not provide obvious, readily detectable targets for exploratory drilling. Their discovery requires thorough and imaginative subsurface geologic investigation, sophisticated interpretation of seismic results, and a willingness to accept a higher dry-hole risk in exploratory drilling. Many of the major onshore discoveries made in the United States and Canada during the past decade are large fields of this type, including *Bell Creek*, *Hilight*, and *Hartzog Draw* in the Powder River Basin and *Yowlumne* in the San Joaquin Valley. The largest pinnacle reef fields, such as the recent discoveries in the *West Pembina* area in the Alberta Basin, could also be included in this category because of their small areal extent, even though they are detectable. Because such fields have been found in some provinces, it is reasonable to assume that they can be found in others with similar geologic histories and characteristics, particularly those producing provinces in which exploratory drilling has been both modest in level and focused entirely on detectable traps. In the author's judgment, possibly another 100 to 200 large fields of this type will be discovered as exploratory drilling becomes more intensive worldwide.*

---

*Halbouty has argued that large amounts of petroleum remain to be discovered in such subtle traps. See, for example, his "Rationale for Deliberate Pursuit of Stratigraphic, Unconformity, and Paleogeomorphic Traps," in R. E. King (ed.), *Stratigraphic Oil and Gas Fields: Classification, Exploration Methods, and Case Histories*, American Association of Petroleum Geologists and Society of Exploration Geophysicists, Tulsa, Okla., 1972, pp. 3-7. The possibility outlined here appears to be the limits of the validity of his argument, because large non-giant fields appear to be the practical maximum size of most types of subtle traps. The 10 to 15 giant fields discovered in subtle traps to date are so small in number that they do not seriously weaken this generalization. (For a description of the various types and likely sizes of stratigraphic traps, of which subtle traps are a subset, see Levorsen, *Geology of Petroleum*, Chap. 7.) For an evaluation of subtle trap possibilities in giant fields, focusing on variations in geologic history and characteristics among provinces, see R. E. King, "Prospects for

The second consideration is the number of large fields that will ultimately be discovered in the Arabian-Iranian province. To date, only a small number have been identified for several reasons. The number of exploratory wells drilled has not been great relative to the potential; the wells that have been drilled have been primarily on structures of super-giant and giant dimensions; and most of the potentially large fields that have been found have not been developed because of their higher costs. Some of the exploratory wells drilled that are listed as suspended or abandoned could also have found large fields with lower productivity wells that were noncommercial by Middle East standards, but would still be economic at production costs of only 10% to 25% of current world oil prices. The extent to which known small structures remain to be drilled in the Middle East could not be determined. But if the ultimate field size distribution in the Middle East proves to be unimodal and highly skewed,* the number of large fields remaining to be discovered will exceed the ultimate number of giant field discoveries in the Middle East (i.e., the number would be at least 150 to 200). However, if nearly all the traps in the Middle East prove to be associated with the large regional structural features on which most of the known fields are located, the number of large fields that will ultimately be discovered there is unlikely to exceed 50 to 100.

With the inclusion of the additional amounts suggested by these two considerations (allowing for some overlap between them), the estimated number of future large non-giant oil field discoveries totals 400 to 800. Assuming that the average-size large non-giant field will contain 150 to 200 million barrels, the amount of oil remaining to be discovered in large non-giant fields is 60 to 160 billion barrels. (The current average of 155 million barrels, derived from Table 3.2, is likely to understate the future average (1) because it excludes possible

---

Large Increases in World Hydrocarbon Reserves from Stratigraphic Traps," *Proceedings, Eighth World Petroleum Congress*, Vol. 2, Applied Science Publishers, London, 1971, pp. 221-226.

*For a discussion of this hypothesis, see G. M. Kaufman, Y. Balcer, and D. Kruyt, "A Probabilistic Model of Oil and Gas Discovery," in Haun (ed.), *Methods of Estimating the Volume of Undiscovered Oil and Gas Resources*, pp. 113-142.

giant fields, most of which contain 250 to 400 million barrels each, that may ultimately prove to be only large non-giant fields and (2) because it is dominated by the many large non-giant fields in the Mississippi and Niger deltas, few of which exceed 200 million barrels.)

The major question mark is the significance of other non-giants for ultimate recovery. The argument has been made that non-giant fields will become increasingly important in proportion to total world oil resources. This argument is based on analogy, comparing the distribution of resources by field size in the United States with that in the rest of the world. The majority (53.1%) of the oil discovered to date in the United States is in non-giants. Most of the exploratory drilling worldwide until now has also occurred in the United States. Assuming a decreasing average size of discovery within a province as exploration progresses (an assumption that has strong empirical and theoretical support) and assuming more extensive drilling worldwide, the argument is made that the distribution worldwide will approach that in the United States.[*]

The hypothesis that smaller fields will become proportionately more important worldwide could be a reasonable one. The assumption that they will be as important in the rest of the world as they are in the United States does not. Several factors suggest otherwise.

The importance of smaller fields in the United States is principally the result not of extensive exploration but of the prevalence of small provinces and of the particular characteristics of the provinces in the Gulf Coast Geosyncline. There are more than 40 non-major producing provinces in this country (nearly all of which have an ultimate recovery of 0.1 to 5.0 billion barrels). Given that the largest field in a province typically has between 10% to 50% of the total resources of the province,[†] such small provinces are unlikely to have an unusually large concentration of their resources in giant fields.

---

[*]For example, such an argument is implied in the work of B. F. Grossling, "The Petroleum Challenge with Respect to the Developing Nations," in Meyer (ed.), *The Future Supply of Nature-Made Petroleum and Gas*, pp. 57-69.

[†]There are a few exceptions to this rule. On the high side, *Prudhoe Bay*, *Ta-ch'ing* (Sung-liao), and *San Ardo* (Salinas) have over

The distribution of known recoverable resources by field size differs considerably among the different groups of provinces within the United States. The major provinces outside the Gulf Coast Geosyncline (the North Slope, San Joaquin, Los Angeles, and Permian basins) have 39.4 billion barrels (72.8%) in known and potential giant fields, 7.9 billion barrels (14.6%) in large fields, and 6.8 billion barrels (12.6%) in smaller fields. The provinces, major and otherwise, of the Gulf Coast Geosyncline (East Texas, Texas Gulf Coast, Mississippi Delta, Arkla, and Mid-Gulf Coast) with their many salt dome structures have 17.3 billion barrels (37.9%) in giant fields, 13.6 billion barrels (29.7%) in large fields, and 14.8 billion barrels (32.4%) in smaller fields. All the other producing provinces have 12.7 billion barrels (26.4%) in giant fields, 13.5 billion barrels (28.0%) in large fields, and 22.0 billion barrels (45.6%) in smaller fields. The distribution by field size in the first four major provinces does not differ significantly from that of most of the other major provinces of the world, despite the fact that all but the North Slope have been extensively explored. The distribution by field size in the small provinces is similar to that in small provinces elsewhere in the world. Smaller fields have provided a larger proportion of total ultimately recoverable oil in the United States because a larger proportion of the total has been found in small provinces.

Furthermore, the importance of smaller fields is not increasing in the United States. Between 1961 and 1976, 18.7 billion barrels of oil were discovered in new fields or in new pools of exploratory significance in the United States.* Of this amount, approximately 5.6 billion barrels were found in fields smaller than 100 million barrels known recovery (crude oil, gas liquids, and oil-equivalent gas). The proportion of the total amount discovered recently in smaller fields

---

90% of the oil discovered to date in their respective provinces. The largest fields in the Mississippi and Niger deltas have little more than 5% of the total resources found to date in these provinces.

*American Gas Association, American Petroleum Institute, and Canadian Petroleum Association, *Reserves of Crude Oil, Natural Gas Liquids, and Natural Gas in the United States and Canada as of December 31, 1976*, Washington, D.C., 1977, p. 25.

(30.0%) is almost identical with the proportion of the total discovered since exploration began in the United States (29.5%). In the future, the proportion of the total in smaller fields may even decrease. Most of the potential for enhanced oil recovery in the United States is concentrated in giant and large fields, particularly in the heavy oil fields of California and the carbonate oil reservoirs of the Permian Basin. The exploitation of enhanced recovery opportunities will both increase the proportion of ultimate recoverable oil in currently recognized giant and large fields and elevate numerous medium-size fields into the large category. Additional discoveries of smaller fields will only partially offset this effect. Because of these trends, the proportion of the oil that will ultimately be produced in the United States from smaller fields will probably range from 25% to 30% of the total. For similar reasons, the proportion in Canada (currently 32.9%) will fall in the same range.

These considerations set some useful limits on the amounts of crude oil to be forthcoming from medium, small, and very small fields worldwide. Assuming that the proportion of total recoverable resources in smaller fields in both the explored provinces and the potential provinces in which smaller fields can be developed outside the United States, Canada, and the Middle East will approach the levels of the Unites States and Canada with further exploratory drilling (i.e., approach 25% to 30% of the total) and that the amount discovered and developed in these provinces in giant and large fields will increase from the current 290 billion barrels to from 375 to 450 billion barrels, the total amounts remaining to be discovered and developed in very small to medium-size fields in these provinces will range from 75 to 116 and 100 to 148 billion barrels.

For a number of reasons, this range appears to be too high. Many of the larger provinces elsewhere in the world have already been explored extensively. There is nothing quite comparable to the greater Gulf Coast Geosyncline, that is, several adjacent major and intermediate-size provinces with large numbers of small fields. The distribution of province sizes (in terms of recoverable resources) and types is less favorable to small fields in the rest of the world. At best, the amount remaining to be discovered in very small to medium-size fields elsewhere in the world is more likely to be only 20% to 25% of the total. Assuming that it will be 22.5%, the total amount remaining to be discovered in smaller fields outside the United States, Canada, and the Middle East is likely to be 60 to 90 billion barrels.

Because of the unique geology of the Middle East and the concentration of its known resources in super-giants, it is most unlikely that the amount of oil discovered in very small to medium-size fields in this area would ever exceed 2% to 3% of its current recoverable resources. Generally speaking, the importance of very small to medium-size fields increases proportionately as province size decreases. Moreover, the aggregate amount of oil found in very small to medium-size fields in any province has never been absolutely large. The four provinces (Permian, Texas Gulf Coast, Mississippi Delta, and Volga-Ural) with the largest amounts of oil in such fields contain 4.5 to 6.0 billion barrels each in fields smaller than 100 million barrels. Reaching 15 billion barrels in the Middle East (3%) would require the discovery of 1000 15-million barrel fields. On the basis of these considerations, an estimate of 6 to 15 billion barrels in smaller fields remaining to be discovered in the Middle East appears to be reasonable.

The total amount remaining to be discovered and developed in smaller fields is estimated here to be 73 to 120 billion barrels (assuming 7 to 15 billion barrels remaining to be discovered in smaller fields in the United States and Canada). This estimate assumes that almost no smaller fields will be developed in the offshore Arctic and deep ocean provinces because they will be both uneconomic to search for (given a declining discovery rate as field size decreases) and uneconomic to develop.

This estimate is reinforced by a consideration of the absolute number of very small to medium-size fields that would have to be discovered to provide significant amounts of recoverable oil resources. Assuming that the average medium-size field contains 50 million barrels, the average small field contains 15 million barrels, and the average very small field contains 4 million barrels, it would take 2000, 6650, and 25,000 discoveries in each, respectively, to add up to 100 billion barrels each. By comparison, despite the high rate of exploratory drilling in the United States in the past 30 years, no more than 25 small and medium-size oil fields have been discovered here in any year since 1946, and the average number of such discoveries is roughly half that amount.

This analysis of future discoveries by size of field suggests that major additions from future discoveries are unlikely. With 30 to 100 billion barrels in super-giants (4 to 10 more fields at 7.5 to 10 billion barrels each), 100 to 175 billion barrels in other giants (125 to 175 more fields at 800 to 1000 million barrels each), 60 to 160 billion barrels in large non-giants (400 to 800 more fields at 150 to 200 million barrels each), and 73 to 120 billion barrels in smaller fields, future discoveries are likely to total 263 to 555 billion barrels.*

RESERVE GROWTH IN KNOWN FIELDS

Estimates of ultimate conventional world oil resources have traditionally concentrated on the amounts remaining to be discovered. The analysis presented in this report of the discovery and development of giant fields and the provinces containing them suggests that this may be an inappropriate focus. If future discoveries follow the same patterns as past discoveries, it appears that 65% to 80% of the oil that will ultimately be discovered has already been discovered. The primary focus of efforts at estimation should thus be additional amounts of oil that may be recovered from known fields.

Estimated total recovery from known fields may increase in several ways. The spatial dimensions of the field may be extended, both horizontally and vertically. More wells may be drilled within the boundaries of the field. Secondary and enhanced recovery operations may be implemented within the field. Although the available data on giant

---

*For two analyses that consider comparable variables and arrive at estimates of future discoveries worldwide in a roughly similar range, see I. H. MacKay and F. K. North, "Undiscovered Oil Reserves," in Haun (ed.), *Methods of Estimating the Volume of Undiscovered Oil and Gas Resources*, pp. 76-86; and Klemme, "World Oil and Gas Reserves."

fields are insufficient to permit a comprehensive evaluation of the potential for increases in total recovery, they do permit several observations.

The potential for reserve growth from extensive development of known fields does not appear to be significant. The horizontal boundaries of nearly all of the known and potential giant oil fields of the world have been determined. The one significant exception would be some of the potential giant oil fields in the Middle East in which only the discovery well has been drilled to date. Although extensions of known fields were a major source of reserve additions in the past when demand often was insufficient to justify an initial investment in complete development, in the future major extensions are apt to occur only within three to five years after the initial discovery. This is particularly true of offshore fields, now the majority of giant discoveries, which have to be fully delineated before the optimal location of the producing platforms can be determined. Discoveries onshore and offshore in countries other than the largest producing countries of the Middle East are also likely to be developed quickly, either to replace imports or to maintain or increase existing export capabilities.

Although new-pool *discoveries* in known fields do not offer much promise of significant additions, development of known pools that currently are not producing could provide significant increases. Most of the giant anticlinal fields of the Middle East have several reservoirs, particularly those in southwest Iran, southern Iraq, Kuwait, Saudi Arabia, and Abu Dhabi. In general, only the least expensive of these have been developed and included in current reserve estimates. The others, although they are more expensive to produce, could be developed at a cost equal to only 10% to 25% of current Middle East prices. However, their quantitative potential is uncertain, for in most cases the only information available about them is that they exist.

Intensive development of known giant fields is considerably more promising. In only three of the 17 major producing countries--Canada, the Soviet Union, and the United States--is a high well density in the major fields prevalent. Three others--Indonesia, Mexico, and Venezuela--have a mixed pattern of well densities. In Abu Dhabi, Algeria, Iran,

Iraq, Kuwait, Libya, the Neutral Zone, Nigeria, Saudi Arabia, and the United Kingdom, the major fields are characterized by a few widely separated, highly productive wells.* As daily productivity per well gradually declines in the giant fields of these countries, drilling them to tighter spacing is likely to increase total recovery, as it has in many of the fields in the United States and the Soviet Union. In many cases, this additional development drilling will be a necessary accompaniment to secondary and enhanced recovery operations in these fields.

The data necessary for a comprehensive assessment of the potential of secondary and enhanced recovery operations worldwide could not be obtained for this report. Comprehensive data on original oil-in-place and current recovery factors could not be obtained for giant fields except for those in Canada, the United States, and Norway. Estimates of original oil-in-place or the necessary information for calculating it for some giant fields in other countries were found in the literature surveyed. Estimates of original oil-in-place by country were also found for several other nations.† Piecing these various sources together suggested that the original oil-in-place in known fields worldwide as of the end of 1975 was between 3.0 and 3.5 trillion barrels. If this range of estimates is correct, worldwide recovery is currently 28.9% to 33.7% of known original oil-in-place.

The information available only permitted a partial analysis of the extent to which the effect of secondary and enhanced recovery operations has already been incorporated in current estimates of total recovery. Of the 272 known giant fields, at least 160 have secondary or enhanced recovery operations planned or under way. These fields contain roughly 70% of the known recoverable oil in giant fields (546 to 776 billion barrels). Of the remaining 112, production had not yet begun in 27 by the end of 1975, and cumulative production was less than 10% of current estimates of total recovery in 23 and between 10%

---

*There are no public data available about the well spacing in China.

†World Energy Conference, *Survey of Energy Resources, 1976*, London, 1977.

and 25% of total recovery in another 21. (In all these circumstances, lack of secondary recovery or pressure maintenance is understandable for economic reasons.) Probably 40% to 60% of the other 50 fields either had a pressure maintenance or secondary recovery project under way or a strong natural water drive that was not indicated by the data gathered for this report.

Some amount of secondary and enhanced recovery is thus already incorporated in the data presented in this report in the text tables and Appendix A. This is particularly true for giant fields in Canada, the Soviet Union, and the United States. To a lesser degree, it is true for the giant fields in Abu Dhabi, Algeria, Indonesia, Iran, Iraq, Kuwait, Libya, Mexico, Saudi Arabia, the United Kingdom, and Venezuela. The extent to which it may be incorporated in the estimates for China, the Neutral Zone, and Nigeria could not be determined.

The estimates of reserves and total recovery provided in Appendix A do not, therefore, indicate the full potential of additional recovery from future investment worldwide. Many of the reserve estimates reflect only the investment made at a very low cost of production (less than $0.50 per barrel). If one were to postulate a willingness to spend considerably higher amounts to produce the remaining oil-in-place (such as $5 to $30 per barrel, the higher figure being a common estimate of the cost at which substitutes for crude oil will become available in substantial quantities), ultimate reserves are likely to be increased substantially.

For an initial approach to estimating the potential for additional recovery worldwide, the experience and potential of the United States may provide a useful analogy. The United States has had nearly 15% of current total world recovery, thus constituting a significant sample. Nearly all of the known fields are fully developed and have been producing for many years. Investment in production facilities in most of the fields has been substantial.

As of the end of 1975, 148 of the 442 billion barrels of the original oil-in-place in the United States either had been produced or were considered to be producible reserves, making the current recovery factor 33.5%. Assuming that world oil prices will rise to at least $30 per barrel in the future (constant dollars), the ultimate recovery factor is estimated to be between 40% and 50%, the width of the range resulting primarily from uncertainties about the effectiveness of enhanced recovery operations and the availability of the materials needed for these operations.* If this same range of recovery is attained elsewhere in the world, additional recovery from known fields worldwide (including the United States) would be 185 to 485 billion barrels (assuming 3.0 trillion barrels known original oil-in-place) or 385 to 735 billion barrels (assuming 3.5 trillion barrels known original oil-in-place). With the most likely range of known original oil-in-place being 3.1 to 3.2 trillion barrels, the additional recovery worldwide suggested by this analogy is 225 to 265 billion barrels (at 40% recovery) and 535 to 585 billion (at 50% recovery).

This analogy, although instructive, could prove to be misleading. There are several reasons for thinking that the U.S. experience is atypical and thus likely to understate the potential elsewhere in the world: (1) Recovery can vary widely from field to field, the variation depending primarily on the viscosity of the oil, the natural drive mechanism in the reservoir, and the characteristics of the reservoir rock. The distribution of these characteristics in oil fields within the United States could be less favorable to high rates of recovery than the distribution of these characteristics in oil fields elsewhere in the world, particularly considering the large proportion of oil-in-place in this country in low-permeability reservoirs in the Permian Basin and in high-viscosity oil fields in California. Although the available data are insufficient to make a precise statistical comparison, the high rates of productivity per well and generally high API gravities of oil production in most of the major producing countries of the world suggest that this is the case. (2) In the United States, current recovery rates are higher in giant fields than in other fields. The proportion of oil in giant fields in the United States is less than

---

*For two recent detailed evaluations of enhanced oil recovery in the United States, see National Petroleum Council, *Enhanced Oil Recovery*, Washington, D.C., 1976; and Office of Technology Assessment, *Enhanced Oil Recovery Potential in the United States*, Washington, D.C., 1977.

it is elsewhere in the world. If recovery rates are higher in giant fields abroad, the total recovery factor will be larger than it is in the United States. (3) Many of the oil fields in the United States were discovered and developed early in the history of the oil industry. Much or all of their production thus occurred during a period of primitive production technology and under institutional circumstances, such as the rule of capture and the absence of unit operations, that resulted in a substantial waste of reservoir energy and consequently lower recovery rates.* In those areas where production began or occurred primarily after the onset of state production regulation, recovery rates are considerably higher.† Because nearly all oil production outside the United States has occurred under more favorable institutional circumstances, higher recovery rates in these countries are likely, all other things being equal.

These three reasons suggest that an approach to estimating additional recovery potential worldwide that takes them into account will yield more reliable estimates of future potential. The following survey of additional recovery potential by region provides such an approach. For each region (and the major producing countries within it), the potential for additional recovery from known fields (in terms of additions to the estimates of Appendix A) is estimated, based on a brief discussion of the relevant crude oil and reservoir characteristics and past production histories.

The ultimate recovery factor in Canada is likely to be similar to the lower half (40% to 45%) of the projected range for the United States.‡ Ultimate recovery will be relatively low because of the

---

*For a description and analysis of the development of and the rationale for petroleum conservation in the United States, see S. L. McDonald, *Petroleum Conservation in the United States: An Economic Analysis*, The Johns Hopkins Press, Baltimore, Md., 1971.

†In the large and giant fields that still have substantial remaining reserves in the East Texas and Texas Gulf Coast provinces, ultimate recovery with enhanced oil recovery will exceed 70% of the original oil-in-place. This is atypical, for the reservoirs in these fields have excellent porosity and permeability, but it shows what can be achieved under favorable circumstances.

‡For a discussion of ultimate recovery in Alberta, see Energy Resources Conservation Board, *Alberta's Reserves of Crude Oil, Gas, Natural Gas Liquids, and Sulphur*, Sec. 9.

predominance of the low-permeability *Pembina* field which has one-sixth of known Canadian original oil-in-place. Additional recovery in Mexico is highly uncertain because it is on the threshold of a massive expansion of reserves from the new discoveries in the Reforma area. The fields discovered in this area to date are reported to have excellent horizontal and vertical permeability as well as an active water drive. With edge water injection and moderate production rates, they are expected to have at least 40% to 45% recovery, and it would not be surprising if the recovery factor reached 50% to 60%. Including the United States, the total additional recovery potential for North America is estimated to be 43 to 95 billion barrels.

The recovery factor in Venezuela is currently less than 25%. With the expansion of secondary recovery operations in the light- and medium-gravity fields and the extensive use of thermal recovery in the heavy oil fields, the amount recovered in known fields should increase by between 25% and 50%. The recovery factor in Argentina, Ecuador, Peru, and Trinidad and Tobago is also low (20% to 25%). With the widespread implementation of secondary recovery operations in the first three countries, the total amount recovered could increase by 75% to 125%. Thermal recovery operations have the most promise in the heavy oil fields of Trinidad and Tobago. Because the recovery factor in Brazil is already estimated to be 40%, the additional potential is limited. Some increase in recovery in Colombia is likely in view of the low returns to producers in the past. On the basis of these considerations, the total additional recovery potential from known fields in South America is estimated at 22 to 40 billion barrels.

The recovery factor in the North Sea fields is currently estimated to be 30% in Norway and 35% to 40% in the United Kingdom. If recovery can be increased in the *Ekofisk* complex, the amount recovered could increase by 40% to 60% in Norway. With widespread water injection and the use of subsea completions, a 20% to 30% increase in the amount recovered would be expected with more favorable economics in the United Kingdom. The potential here is uncertain, however, because of the unprecedented combination of high initial recovery factors and the projected rapid depletion rates. Thermal recovery in the heavy oil fields of West Germany and The Netherlands promises to increase the total amount recovered in these two countries by 25% to 50%. The additional recovery potential from known fields in Western Europe is thus limited to 5 to 10 billion barrels.

The average recovery factor in the Soviet Union is estimated at 45%; however, this may represent plans more than realizations.* Extensive implementation of enhanced oil recovery operations is planned for the Soviet Union by the year 2000. Because of the many reservoir problems created by Soviet waterflooding practice, the total amount recovered is not likely to increase by more than 20% to 40%. The prospects for Rumania, which has similar production practices, are comparable. Additional recovery prospects for Eastern Europe and the Soviet Union thus total 20 to 40 billion barrels.

The recovery factor in the North African countries is an estimated 35% to 40%. In some fields of Libya, such as *Sarir, Intisar A,* and *Intisar D,* the recovery factor is already estimated to be 50% to 60%. No indications of secondary recovery could be found for the other major fields. Most of the larger fields in Algeria, Egypt, and Tunisia already have additional recovery operations planned or under way. Enhanced recovery operations similar to the one planned for *Hassi Messaoud* in Algeria promise major increases. Overall, the range of potential increases in the total amount recovered from known fields in North Africa appears to be 20% to 40%. Because most of the fields in Nigeria have good sandstone reservoirs with natural water drives, current recovery factors are already high (40% to 50%). Like the geologically similar Mississippi Delta, increases in the total amount recovered in the Niger Delta will probably be no more than 15% to 30%. Production in Angola, the Congo, and Gabon is characterized by high costs and low recovery rates. Improved economics could increase the total amount recovered by 40% to 60%. In all the African countries, the additional recovery potential is estimated to be 15 to 30 billion barrels.

---

*An estimate of the recovery factor in the Soviet Union is provided by N. S. Erofeev, M. L. Surguchev, and E. M. Halimov, "Enhanced Oil Recovery Is an Important Way To Increase Oil Resources," in Meyer (ed.), *The Future Supply of Nature-Made Petroleum and Gas*, pp. 425-438. For a discussion of some of the problems associated with Soviet recovery practice, see Owen, *Trek of the Oil Finders*, pp. 1402-1410.

The Middle East appears to have intriguing possibilities for additional recovery. Most current estimates of total recovery are based on very low levels of investment relative to the potential. Most of the major fields, particularly the super-giants, have good to excellent reservoir characteristics (as evidenced by the continued high daily rates of production per well). Annual production rates are low, not exceeding 2% to 3% of ultimate recovery in nearly all cases. The governments are strongly aware of the importance of good conservation practices and do not face strong financial incentives to ignore them.

However, any consideration of the potential of enhanced recovery in the Middle East needs to differentiate among the different groups of producing fields. Because the heavily fissured and fractured limestone fields of southwest Iran and northern Iraq have low matrix permeabilities, their recovery potential is limited. However, with a long producing life (75 to 100 more years), a major investment in desalting facilities, and successful secondary recovery (currently waterflooding in *Kirkuk* and planned gas injection in southwest Iran), the amount recovered could increase 50% from present levels. Nearly all of the sandstone reservoirs of southern Iraq, Kuwait, the Neutral Zone, and the northern half of the Persian/Arabian Gulf have excellent porosity and permeability. With peripheral water injection, long production lifetimes, investment in desalters, infill drilling, and a willingness to produce fluids up to 90% to 95% water cuts, these reservoirs should ultimately have recovery factors similar to those of the major fields of East Texas and the Texas Gulf Coast. Thermal recovery has possibilities in the heavy oil reservoirs in the fields of this area, some of which are still undeveloped, as well as in the heavy oil fields of northern Iraq and Syria. Most of the major limestone reservoirs in Abu Dhabi and onshore Saudi Arabia have good to excellent porosity and permeability. They too should yield high recovery factors with peripheral water injection, long production lifetimes, investment in desalters, infill drilling, and a willingness to produce with high water cuts.* Additional recovery potential in the other countries of

---

*A hint of the potential of waterflooding in Saudi Arabia is given in F. W. Dolce and J. H. Rebold, "Observation of Oil-Water Contact

the Arabian Peninsula (Bahrein, Dubai, Oman, and Qatar) is limited, but some increases will occur, particularly in Qatar. This assessment suggests that additional recovery in known fields in the Middle East should be 250 to 400 billion barrels, possibly more with highly successful waterflooding.* The effort to achieve this will require hundreds of billions of dollars in investment and operating costs over several decades, but this amount is only a small fraction of the costs of alternative energy supplies of similar magnitude.

Information about the additional recovery potential of Asia/Oceania is limited. From all indications, production practice in China follows the Soviet model, with waterflooding beginning early in the life of the field. Implementation of secondary recovery operations in Indonesia depends primarily on economics; in central Sumatra, in particular, there is a definite potential for further additions. Most of the major fields in Brunei, India, and Malaysia are at an early stage of their productive life, making secondary recovery a definite possibility. The recovery factor in Australia is already high (45% to 50%), limiting the potential for further increases. Considering the various uncertainties, the additional recovery potential in known fields in Asia/Oceania is most likely on the order of 15 to 25 billion barrels.

The sum of the estimates of additional recovery by region is 370 to 640 billion barrels. This range is somewhat higher than that yielded by the analogical approach, particularly at the lower end. With a favorable resolution of the current uncertainties about original oil-in-place and likely displacement efficiencies in the Middle East, additional recovery could be even higher. The combined potential for reserve growth in known fields from extensions, new-pool discoveries, full development, and additional recovery is estimated to range from 420 to 730 billion barrels.

Advance and Determination of Oil Displacement Efficiency by Water," *Proceedings, Fifth Arab Petroleum Congress*, Cairo, 1965. The authors reported that in one observation well on the flank of the *Abqaiq* field displacement efficiency was reaching 75% to 85% at the bottom of the waterflood front, even though an equilibrium residual oil saturation had not yet been reached.

*For a similar assessment of Middle East potential, see Beydoun and Dunnington, *The Petroleum Geology and Resources of the Middle East*, pp. 80-84.

ULTIMATE WORLD OIL RESOURCES

The ultimate recoverable conventional crude oil resources of the world are estimated to be 1700 to 2300 billion barrels.* This includes 335 billion barrels that had already been produced by the end of 1975, 676 billion barrels of proved and probable reserves, from 263 to 555 billion barrels to be added from future discoveries, and from 420 to 730 billion barrels to be added from further development of and additional recovery from known fields. Additional recovery is thus projected to provide 57% to 61% of future additions.

The range of 1700 to 2300 billion barrels incorporates most of the major uncertainties in world oil resource estimation, assuming that the exploration of unexplored provinces will yield results basically analogous to those in the explored provinces and that future development and recovery worldwide will be basically comparable to that likely to be achieved in the most highly developed fields today. If the ultimate amount proves to be outside this range, one or more non-analogous events or developments will have to occur.

Ultimate conventional crude oil resources will be less than 1700 billion barrels only if there is a sharp drop-off in discoveries of giant and large fields (most likely from a lack of new major provinces), and if additional recovery efforts yield poor results, even at significantly higher prices. For ultimate conventional crude oil resources to be more than 2300 billion barrels, one of the following developments will have to occur: (1) proportionately more of the unexplored provinces prove to be major provinces than have the explored provinces; (2) a few new major provinces, larger than any known province other than the Arabian-Iranian Basin (i.e., 50 to 100 billion barrels and several super-giants each), will be opened up; (3) one or more *Ghawar*-size

*Although the components do not quite equal the totals, the estimate is rounded off because the uncertainty in estimation does not permit any meaningful numbers beyond two significant digits. Readers familiar with recent estimates of world oil resources will notice that the midpoint of the range coincides with the "consensus" estimate of 2000 billion barrels. Contrary to appearances, the agreement is purely fortuitous, for the totals presented here were not determined until the disaggregated analysis was completed.

fields will be discovered; or (4) additional recovery efforts, particularly in the Middle East, prove to be highly successful.

None of these developments are impossible. It appears to be most unlikely, however, that ultimate conventional oil resources would be less than 1700 billion barrels. The known prospects in the current producing provinces and the record of continued giant field discoveries since 1975 suggest that a very sharp drop-off in discoveries is improbable. The prospects for secondary recovery alone, a technique that has been used successfully worldwide in many geologic circumstances, are substantial, particularly in the Middle East.

Reducing the uncertainty at the high end of the estimate cannot be so easily substantiated. Each of the three logical possibilities for future major discoveries would be a non-analogous event, given the past history of exploration. They are excluded from the range estimated here because they are non-analogous and because a sketchy survey of the known geology of the unexplored and lightly explored provinces did not indicate a highly promising potential. However, because exploratory surprises do occur, the most recent being the opening up of the Reforma area in Mexico, each of these possibilities should be kept in mind as exploration continues. Further exploratory activity, particularly in the next 10 to 20 years, will provide the information to indicate whether any of them are realistic. Very successful recovery efforts resulting from high rates of investment and improvements in recovery technology appear to have a reasonable probability of occurring. They are excluded from the range estimated here because they fall outside the realm of present experience. If they do occur, their primary effect will be on oil availability in the next century.

The distribution by field size of ultimate crude oil resources is expected to change only slightly from the present distribution. (An approximate ultimate distribution by field size is provided in Table 5.1.*) Super-giant fields will become somewhat less important; large non-giant fields will become somewhat more important. Other giant,

---

*The number of fields in each category reflects both new discoveries and transfers from other categories because of reserve growth.

Table 5.1

ESTIMATED DISTRIBUTION OF ULTIMATE RECOVERABLE OIL RESOURCES OF THE WORLD BY FIELD SIZE

Field Size	Number	Amount (In billions of barrels)
Super-giants	45-55	825-1065
Other giants	475-550	475-635
Large non-giants	1000-1400	200-350
Other non-giants	No estimate	200-250
Total		1700-2300

large non-giant, and other non-giant fields are predicted to comprise a greater proportion of future discoveries than they have in the past. The share of additional recovery from super-giant fields--known, potential, and prospective--is predicted to be greater than the current proportion of known recoverable resources in super-giant fields, primarily because of additional recovery from the super-giant fields of the Middle East.

The geographic distribution by region (including the adjacent continental margins) of ultimate conventional world crude oil resources is projected to change only slightly. (An approximate distribution by region is provided in Table 5.2.) Despite the high promise of the Reforma area of Mexico, the importance of North America is projected to decline slightly. The overall proportions in South America, Eastern Europe/the Soviet Union, Africa, and the Middle East are expected to remain basically unchanged. The respective shares in Western Europe and Asia/Oceania (including Antarctica) are expected to increase slightly. Considering that these disaggregated projections are somewhat more uncertain than the range of the total, these differences are basically insignificant.

The geographic distribution projected here differs significantly from those of most other estimates with similar totals. In particular,

Table 5.2

ESTIMATED ULTIMATE CONVENTIONAL WORLD CRUDE OIL RESOURCES BY REGION

(In billions of barrels)

Region	Known	Potential	Total
North America	179.8	100-200	280-380
South America	68.4	52-92	120-160
Western Europe	24.6	25-45	50-70
Eastern Europe/ Soviet Union	102.4	63-123	165-225
Africa	75.6	45-94	120-170
Middle East	509.9	350-630	860-1140
Asia/Oceania	50.8	54-104	105-155
Total	1011.5	688-1288	1700-2300

other estimates* project considerably more for the Soviet Union (twice as much in some cases), somewhat more for Western Europe and Asia/ Oceania, and much less for the Middle East. The estimates for North America, South America, and Africa in Table 5.2 and in other estimates are roughly similar, given the prevailing uncertainty. The primary reasons for these differences are (1) other estimates have not explicitly incorporated an assessment of the additional recovery potential of the Middle East, and (2) other estimates overstate the oil potential of the relatively unexplored provinces in the Soviet Union, not taking into account the fact that those that are beginning to be explored are turning out to be principally gas provinces.

OIL RESOURCES AND OIL AVAILABILITY

The preceding review of world oil resources suggests that there will ultimately be 1700 to 2300 billion barrels of oil recovered from conventional crude oil fields. This range, less cumulative production,

---

*For example, Moody and Esser, "An Estimate of the World's Recoverable Crude Oil Resource" and the consensus estimate published by the Conservation Commission World Energy Conference, *Report on Oil Resources*.

is equivalent to 60 to 90 years of world consumption at current rates. However, it does not include the hundreds of billions of barrels that could become available from heavy oil (tar) sand and oil shale deposits.

Several factors will determine the rate at which the ultimate resource will become and be available. Because the ultimate resource is concentrated in giant and super-giant fields, the factors determining world oil availability essentially revolve around giant fields: the rate at which they can and will be discovered, the rate at which they will be developed and produced, and their characteristics and economics.

For many of the Arctic and oceanic provinces, the necessary exploration or production technology is not yet available and is unlikely to be so for at least another decade. The technology necessary for exploration and production offshore in Arctic regions (except for shallow waters) is currently in the conceptual and developmental stages in the United States and Canada. Its application in the Soviet Union is in the even more distant future. Although exploratory drilling is now possible in waters 300 to 1500 meters deep, the technology for production in waters of this depth is still in the conceptual and developmental stages. Thus, even if giant fields are discovered in some of the most promising remaining unexplored provinces, production is not likely to be technically possible before 1990. In other areas, such as the offshore areas of China, the necessary technology for exploration and production exists, but may not be employed rapidly for a number of political reasons (although there are recent indications that Chinese oil policy may be moving toward a more rapid development). The implication of the unavailability and the lack of full utilization of the necessary technology is that world oil availability between now and at least 1990 will be determined almost entirely by the availability of oil from known and yet-to-be-discovered fields in known producing provinces.

Because production between now and 1990 will come almost entirely from the known producing provinces, near- and medium-term oil availability will depend primarily on the production policies of the OPEC countries. The current world potential for higher production is concentrated in these countries, particularly Abu Dhabi, Iraq, Kuwait, and Saudi Arabia. The extent to which it will be developed and used depends, first, on whether these countries will undertake or permit the necessary investment in additional productive capacity, the cost of which in many cases will be an order of magnitude or more higher than most of the capacity developed in these countries to date (but one that still provides a complete payback on investment at present world oil prices within one to two years at most). A key factor is the point at which investments in secondary and enhanced recovery operations in giant fields will become desirable in the major exporting countries. Second, the development of additional capacity depends on the extent to which these countries, particularly Saudi Arabia and Abu Dhabi, will want to determine the world crude oil prices. The availability of excess capacity, which they essentially control, will be the principal factor in determining the rate of increase in world oil prices. It will not, however, be the only factor in determining world oil prices. The potential that could be discovered and developed outside the OPEC countries is sufficient to constrain prices for the next 10 to 15 years, assuming that demand does not grow by more than 1.0% to 1.5% annually. Finally, additional capacity will depend on their decisions as to the rate of depletion that they consider to be desirable for their long-run economic future. Oil production in all of these countries is concentrated in a handful of giant and super-giant fields. If no major fields or even if only a few more major fields are discovered in each, these countries will know that the rapid depletion of their existing fields means the depletion of their principal natural economic asset. Their decisions to increase production capacity are thus inescapably decisions about their long-run pattern of economic development.

Future oil availability will also be determined by the economic returns available to both producers and refiners. The period in which incremental oil production capacity could be added very cheaply, a period that was synonymous with the discovery and initial development of the super-giants, is past. Future increments to capacity will come from fields discovered in expensive environments, from fields with lower well productivity, from smaller fields, or from the application of secondary and enhanced recovery. In those countries where the investment in incremental capacity is made by the oil companies rather than by the government, the returns to companies will have to be sufficient if the necessary investments are to be made. Uncertainty, especially about future taxes and royalties, could be a major deterrent to further investment. Moreover, additional investments will be required in the refinery sector. A considerable amount of the undeveloped known potential productive capacity of the world is in heavy, high-sulfur oil fields. Altering existing refinery capacity to process a growing proportion of crude oils of this type will require major investments (although the investment per barrel of capacity is generally substantially less than the investment per daily barrel of enhanced oil recovery capacity). The refinery investment problem is magnified by the trends in the major oil-consuming countries toward implementing stricter sulfur emission standards and toward deemphasizing the use of oil in industrial boilers and for electrical generation. Producing a higher proportion of lighter, lower sulfur products from increasingly heavier, higher sulfur crude oils will require major changes in existing world refinery capacity.

This report has described the importance of giant oil fields to known recoverable conventional world oil resources and has shown the utility of a description and analysis of their distribution and patterns of discovery in assessing ultimate conventional world oil resources. A description and analysis of their economic and geologic characteristics and their political control (which has only been briefly outlined here) would also provide the basis for understanding the factors that will shape the future availability of oil. The fact that world oil resources are concentrated in a relatively small number of giant fields provides the fundamental starting point and the essential clue to understanding the dynamics of the world oil market. Because they occupy such a central role, monitoring future giant field discoveries and development offers the most fruitful means of reducing present uncertainties about ultimate world oil resources.

REFERENCES

Adams, T. D., and M. A. Kirkby, "Estimate of World Gas Reserves," *Proceedings, Ninth World Petroleum Congress*, Vol. 3 (Exploration and Transportation), Applied Science Publishers, London, 1975, pp. 3-9.

Adelman, M. A., *The World Petroleum Market*, The Johns Hopkins Press, Baltimore, Md., 1972.

American Gas Association, American Petroleum Institute, and Canadian Petroleum Association, *Reserves of Crude Oil, Natural Gas Liquids, and Natural Gas in the United States and Canada as of December 31, 1976*, Washington, D.C., 1977.

Beydoun, Z. R., and H. V. Dunnington, *The Petroleum Geology and Resources of the Middle East*, Scientific Press, Beaconsfield, England, 1975.

Burke, R. J., and F. J. Gardner, "The World's Monster Oil Fields and How They Rank," *Oil and Gas Journal*, Vol. 67, No. 2, January 13, 1969.

Childs, O. E., "Implications for Future Petroleum Exploration," in R. F. Meyer (ed.), *The Future Supply of Nature-Made Petroleum and Gas*, Pergamon Press, New York, 1977, pp. 81-99.

Demaison, G. J., "Tar Sands and Supergiant Oil Fields," *American Association of Petroleum Geologists Bulletin*, Vol. 61, No. 11, November 1977, pp. 1950-1961.

Dietz, R. S., and J. C. Holden, "The Breakup of Pangea," *Scientific American*, Vol. 223, No. 5, November 1970.

Dolce, F. W., and J. H. Rebold, "Observation of Oil-Water Contact Advance and Determination of Oil Displacement Efficiency by Water," *Proceedings, Fifth Arab Petroleum Congress*, Cairo, 1965.

Energy Resources Conservation Board, *Alberta's Reserves of Crude Oil, Gas, Natural Gas Liquids, and Sulphur at December 31, 1976*, Calgary, Alberta, 1977.

Erofeev, N. S., M. L. Surguchev, and E. M. Halimov, "Enhanced Oil Recovery Is an Important Way To Increase Oil Resources," in R. F. Meyer (ed.), *The Future Supply of Nature-Made Petroleum and Gas*, Pergamon Press, New York, 1977, pp. 425-438.

Gess, G., and C. Bois, "Study of Petroleum Zones: A Contribution to the Appraisal of Hydrocarbon Resources," in R. F. Meyer (ed.), *The Future Supply of Nature-Made Petroleum and Gas*, Pergamon Press, New York, 1977, pp. 155-178.

"Giant Fields Still Yield Most U.S. Oil," *Oil and Gas Journal*, Vol. 76, No. 5, January 30, 1978.

Grossling, B. F., "The Petroleum Challenge with Respect to the Developing Nations," in R. F. Meyer (ed.), *The Future Supply of Nature-Made Petroleum and Gas*, Pergamon Press, New York, 1977, pp. 57-69.

Halbouty, M. T., "Giant Oil and Gas Fields in the United States," in M. T. Halbouty (ed.), *Geology of Giant Petroleum Fields*, American Association of Petroleum Geologists Memoir 14, Tulsa, Okla., 1970.

------, "Rationale for Deliberate Pursuit of Stratigraphic, Unconformity, and Paleogeomorphic Traps," in R. E. King (ed.), *Stratigraphic Oil and Gas Fields: Classification, Exploration Methods, and Case Histories*, American Association of Petroleum Geologists and Society of Exploration Geophysicists, Tulsa, Okla., 1972, pp. 3-7.

------ (ed.), *Geology of Giant Petroleum Fields*, American Association of Petroleum Geologists Memoir 14, Tulsa, Okla., 1970.

------, et al., "World's Giant Oil and Gas Fields, Geologic Factors Affecting Their Formation and Basin Classification," in M. T. Halbouty (ed.), *Geology of Giant Petroleum Fields*, American Association of Petroleum Geologists Memoir 14, Tulsa, Okla., 1970, pp. 502-555.

Haun, J. D., "Methods of Estimating the Volume of Undiscovered Oil and Gas Resources--AAPG Research Conference," in J. D. Haun (ed.), *Methods of Estimating the Volume of Undiscovered Oil and Gas Resources*, American Association of Petroleum Geologists Studies in Geology No. 1, Tulsa, Okla., 1975, pp. 1-7.

Hobson, G. D., and E. N. Tiratsoo, *Introduction to Petroleum Geology*, Scientific Press, Beaconsfield, England, 1975.

Holmgren, D. A., J. D. Moody, and H. H. Emmerich, "The Structural Settings for Giant Oil and Gas Fields," *Proceedings, Ninth World Petroleum Congress*, Vol. 2 (Exploration and Transportation), Applied Science Publishers, London, 1975, pp. 45-54.

International Petroleum Encyclopedia, Petroleum Publishing Company, Tulsa, Okla., annual.

Ivanhoe, L. F., "Evaluating Prospective Basins," *Oil and Gas Journal*, Vol. 74, Nos. 49, 50, and 51, December 6, 13, and 20, 1976.

Kaufman, G. M., Y. Balcer, and D. Kruyt, "A Probabilistic Model of Oil and Gas Discovery," in J. D. Haun (ed.), *Methods of Estimating the Volume of Undiscovered Oil and Gas Resources*, American Association of Petroleum Geologists Studies in Geology No. 1, Tulsa, Okla., 1975, pp. 113-142.

King, R. E., "Prospects for Large Increases in World Hydrocarbon Reserves from Stratigraphic Traps," *Proceedings, Eighth World Petroleum Congress*, Vol. 2, Applied Science Publishers, London, 1971, pp. 221-226.

Klemme, H. D., "Giant Fields Contain Less Than 1% of World's Fields But 75% of Reserves," *Oil and Gas Journal*, Vol. 75, No. 10, March 7, 1977.

------, "Giant Oil Fields Related to Their Geologic Setting: A Possible Guide to Exploration," *Bulletin of Canadian Petroleum Geology*, Vol. 23, No. 1, March 1975, pp. 30-66.

------, "The Giants and the Supergiants," *Oil and Gas Journal*, Vol. 69, Nos. 9, 10, and 11, March 1, 8, and 15, 1971.

------, "One-Fifth of Reserves Lie Offshore," *Oil and Gas Journal, Petroleum/2000*, Vol. 75, No. 35, August 1977.

------, "Structure-Related Traps Expected To Dominate World Reserve Statistics," *Oil and Gas Journal*, Vol. 71, No. 53, December 31, 1973; and Vol. 72, No. 1, January 7, 1974.

------, "Trends in Basin Development: Possible Economic Implications," *World Petroleum*, October 1971.

------, "World Oil and Gas Reserves from Analysis of Giant Fields and Petroleum Basins (Provinces)," in R. F. Meyer (ed.), *The Future Supply of Nature-Made Petroleum and Gas*, Pergamon Press, New York, 1977, pp. 217-260.

Knebel, G. M., and G. Rodriguez-Eraso, "Habitat of Some Oil," *American Association of Petroleum Geologists Bulletin*, Vol. 40, No. 4, April 1956, pp. 547-561.

Levorsen, A. I., *Geology of Petroleum*, 2d ed., W. H. Freeman and Company, San Francisco, 1967.

Lovejoy, W. F., and P. T. Homan, *Methods of Estimating Reserves of Crude Oil, Natural Gas, and Natural Gas Liquids*, The Johns Hopkins Press, Baltimore, Md., 1965.

MacKay, I. H., and F. K. North, "Undiscovered Oil Reserves," in J. D. Haun (ed.), *Methods of Estimating the Volume of Undiscovered Oil and Gas Resources*, American Association of Petroleum Geologists Studies in Geology No. 1, Tulsa, Okla., 1975, pp. 76-86.

McDonald, S. L., *Petroleum Conservation in the United States: An Economic Analysis*, The Johns Hopkins Press, Baltimore, Md., 1971.

Menard, H. W., and G. Sharman, "Scientific Uses of Random Drilling Models," *Science*, Vol. 190, No. 4212, October 24, 1975, pp. 337-343.

Meyer, R. F., "Geologic Provinces Code Map for Computer Use," *American Association of Petroleum Geologists Bulletin*, Vol. 54, No. 7, July 1970, pp. 1301-1305.

------ (ed.), *The Future Supply of Nature-Made Petroleum and Gas*, Pergamon Press, New York, 1977.

Meyerhoff, A. A., "Economic Impact and Geopolitical Implications of Giant Petroleum Fields," *American Scientist*, Vol. 64, No. 5, September-October 1976, pp. 536-541.

Moody, J. D., "Distribution and Geological Characteristics of Giant Oil Fields," in A. G. Fischer and S. Judson (eds.), *Petroleum and Global Tectonics*, Princeton University Press, Princeton, N.J., 1975, pp. 307-320.

------, and H. H. Emmerich, "Giant Oil Fields of the World," *Proceedings, Twenty-Fourth International Geologic Congress, Section 5, Mineral Fuels*, Montreal, 1972, pp. 161-167.

------, and R. W. Esser, "An Estimate of the World's Recoverable Crude Oil Resource," *Proceedings, Ninth World Petroleum Congress*, Vol. 3 (Exploration and Transportation), Applied Science Publishers, London, 1975, pp. 11-20.

------, J. W. Mooney, and J. Spivak, "Giant Oil Fields of North America," in M. T. Halbouty (ed.), *Geology of Giant Petroleum Fields*, American Association of Petroleum Geologists Memoir 14, Tulsa, Okla., 1970, pp. 8-17.

National Petroleum Council, *Enhanced Oil Recovery*, Washington, D.C., 1976.

Odell, P. R., and K. L. Rosing, *The North Sea Oil Province*, Kogan Page, London, 1975.

Owen, E. W., *Trek of the Oil Finders: A History of Exploration for Petroleum*, American Association of Petroleum Geologists Memoir 6, Tulsa, Okla., 1975.

Phizackerley, P. H., and L. O. Scott, "Major Tar Sand Deposits of the World," *Proceedings, Seventh World Petroleum Congress*, Vol. 3 (Drilling and Production), Elsevier, Barking, England, 1967, pp. 551-571.

Seidl, R. G., "Implications of Changing Oil Prices on Resource Evaluations," in R. F. Meyer (ed.), *The Future Supply of Nature-Made Petroleum and Gas*, Pergamon Press, New York, 1977, pp. 113-137.

Tiratsoo, E. N., *Oilfields of the World*, Scientific Press, Beaconsfield, England, 1973.

U.S. Congress, Office of Technology Assessment, *Enhanced Oil Recovery Potential in the United States*, Washington, D.C., 1977.

World Energy Conference, *Survey of Energy Resources, 1976*, London, 1977.

------, Conservation Commission, *Report on Oil Resources: 1985-2020, Executive Summary*, Istanbul, 1977.

Appendix A

GIANT OIL FIELDS BY COUNTRY AND REGION

INTRODUCTION

This Appendix provides a country-by-country listing of the giant oil fields and recoverable oil resources of the world as of December 31, 1975. Information for the 67 countries that have at least 10 million barrels of known recoverable oil resources each (roughly 0.001% of the known recoverable oil resources of the world) are provided. Because Cameroun and the Philippines attained this status after 1975, they are excluded from this compilation. Summaries of the distribution of known recoverable oil resources by field classification are also provided for seven regions: North America, South America, Western Europe, Eastern Europe and the Soviet Union, Africa, the Middle East, and Asia/Oceania.

The same format is used in each of the country tables. Giant, combination giant, probable giant, and possible giant fields (if any) are listed individually, those in the former two categories in order of decreasing size and those in the latter two alphabetically. The tables provide the discovery year of these fields, their 1975 production, cumulative production to the end of 1975, proved and probable reserves at the end of 1975, estimated total recovery at the end of 1975 (the sum of cumulative production and reserves), a judgment about the accuracy of this estimate and the possible direction of change, and the range of estimates of total recovery appearing in the literature on giant fields since 1970. Production, reserve, and total recovery data are also provided for other fields and for each country as a whole. If there have been any major discoveries since the end of 1975, they are indicated as well. Brief explanatory notes about the overall quality of the data, the individual known and potential giant fields, and the country as a whole are included.

The following notation is used in the tables. Partially offshore fields are indicated by a single asterisk (*). Wholly offshore fields are indicated by a double asterisk (**). Where production has not yet begun, the absence of production is indicated by a dash (--). The lack of any reasonable data on reserves and total recovery is indicated by a blank space. When dependable information about 1975 and cumulative production was not available, the estimated production data are identified by a capital E. Uncertainty in the estimates of total recovery is also indicated where applicable. Any estimate with a reasonable probability of at least a ± 10% change during the next ten years is indicated by a question mark. If the preponderance of the available evidence suggests that an estimate is likely to be understated by at least 10%, the designation "low?" is used. Where an estimate is likely to be overstated by at least 10%, the designation "high?" is used. Where summary reserve and total recovery data are understated because of incomplete information, a plus sign (+) is used to indicate that the total is higher to some unknown degree. In the regional summaries, a minus sign (-) is occasionally used for the data on other fields to indicate that some proportion of the amounts shown belong in the possible-giants category, but that information indicating how much belonged in this category was unavailable. All totals are rounded.

Sources

The general sources used in the preparation of the following tables are listed below. With only a few exceptions, systematic examination of the periodicals shown was limited to the issues of the past 15 years. Region- and country-specific sources (if any) are listed in the appropriate table.

Books and Articles

1. J. P. Albers et al., *Summary Petroleum and Selected Mineral Statistics for 120 Countries, Including Offshore Areas,* Geological Survey Professional Paper 817, Washington, D.C., 1973.
2. American Petroleum Institute, *Petroleum Facts and Figures,* Washington, D.C., 1971.
3. R. J. Burke and F. J. Gardner, "The World's Monster Oil Fields, and How They Rank," *Oil and Gas Journal,* Vol. 67, No. 2, January 13, 1969.
4. Central Intelligence Agency, *Major Oil and Gas Fields of the Free World,* ER 76-10001, January 1976.
5. G. J. Demaison, "Tar Sands and Supergiant Oil Fields," *American Association of Petroleum Geologists Bulletin,* Vol. 61, No. 11, November 1977, pp. 1950-1961.
6. M. T. Halbouty, ed., *Geology of Giant Petroleum Fields,* American Association of Petroleum Geologists Memoir 14, Tulsa, Okla., 1970 (particularly M. T. Halbouty et al., "World's Giant Oil and Gas Fields, Geologic Factors Affecting Their Formation, and Basin Classification," pp. 502-555).
7. D. A. Holmgren, J. D. Moody, and H. H. Emmerich, "The Structural Settings for Giant Oil and Gas Fields," *Proceedings, Ninth World Petroleum Congress,* Vol. 3 (Exploration and Transportation), Applied Science Publishers, London, 1975, pp. 45-54.
8. H. D. Klemme, "Giant Fields Contain Less Than 1% of World's Fields But 75% of Reserves," *Oil and Gas Journal,* Vol. 75, No. 10, March 7, 1977.
9. H. D. Klemme, "Giant Oil Fields Related to Their Geologic Setting: A Possible Guide to Exploration," *Bulletin of Canadian Petroleum Geology,* Vol. 23, No. 1, March 1975, pp. 30-66.
10. H. D. Klemme, "The Giants and the Supergiants," *Oil and Gas Journal,* Vol. 69, Nos. 9, 10, and 11, March 1, 8, and 15, 1971.
11. H. D. Klemme, "One-Fifth of Reserves Lie Offshore," *Oil and Gas Journal, Petroleum/2000,* Vol. 75, No. 35, August 1977, pp. 108-128.
12. H. D. Klemme, "Structure Related Traps Expected To Dominate World-Reserve Statistics," *Oil and Gas Journal,* Vol. 71, No. 53, December 31, 1973, and Vol. 72, No. 1, January 7, 1974.
13. H. D. Klemme, "Trends in Basin Development: Possible Economic Implications," *World Petroleum,* October 1971.
14. G. M. Knebel and G. Rodriguez/Eraso, "Habitat of Some Oil," *American Association of Petroleum Geologists Bulletin,* Vol. 40, No. 4, April 1956, pp. 547-561.
15. J. McCaslin, "The Search Goes On for Giant Fields," *Oil and Gas Journal, Petroleum/2000,* Vol. 75, No. 35, August 1977, pp. 99-106.
16. A. A. Meyerhoff, "Economic Impact and Geopolitical Implications of Giant Petroleum Fields," *American Scientist,* Vol. 64, No. 5, September-October 1976, pp. 536-541.
17. J. D. Moody and R. W. Esser, "An Estimate of the World's Recoverable Crude Oil Resource," *Proceedings, Ninth World Petroleum Congress,* Vol. 3 (Exploration and Transportation), Applied Science Publishers, London, 1975, pp. 11-20.
18. E. W. Owen, *Trek of the Oil Finders: A History of Exploration for Petroleum,* American Association of Petroleum Geologists Memoir 6, Tulsa, Okla., 1975.
19. E. N. Tiratsoo, *Oilfields of the World,* Scientific Press, Beaconsfield, England, 1973.
20. World Energy Conference, *Survey of Energy Resources, 1976,* London, 1977.

Periodicals

1. *American Association of Petroleum Geologists Bulletin* (monthly).
2. *Arab Oil and Gas* (fortnightly).
3. International Oil Scouts Association, *International Oil and Gas Development* (annual).
4. *International Petroleum Encyclopedia* (annual).
5. *Offshore* (monthly).
6. *Oil and Gas Journal* (weekly).
7. *Petroleum Economist* (monthly).
8. *Petroleum Engineer* (monthly).
9. *Petroleum Intelligence Weekly* (weekly).
10. *World Oil* (monthly).
11. *World Petroleum* (monthly).
12. *World Petroleum Report* (annual).

Table A.1

KNOWN RECOVERABLE CRUDE OIL RESOURCES OF NORTH AMERICA BY FIELD CLASSIFICATION AS OF DECEMBER 31, 1975

(In millions of barrels)

Field Classification	Number of Fields	1975 Production	Cumulative Production	Reserves	Total Recovery
Giant fields	55	1,252.2	37,394	34,301	71,695
Combination giants	21	225.9	6,147	2,218	8,365
Probable giants	7	72.1	1,820	880	2,700
Possible giants	21	152.8	4,356	1,319	5,675
Other fields		1,947.6	72,271	19,059	91,330
Total		3,650.6	121,988	57,777	179,765

NOTES: Giant fields are less important in North America than in any other oil-producing region in the world. The 76 known giant fields contain 80.06 billion barrels, only 44.5% of the crude oil discovered and made recoverable to the end of 1975. Another 4.7% (8.38 billion barrels) are in potential giant oil fields. Over half (50.8%) of known recoverable crude oil in North America is in non-giant fields. Approximately 38.05 billion barrels of this are found in 255 large fields (100-500 million barrels ultimate recovery). The remainder is found in small- and medium-size fields concentrated in the United States and Canada.

Giant fields will become more important for North American crude oil production in the future. Although they contain less than half of the total known recoverable resource, they contain roughly two-thirds of known reserves. The known giants have reserves of 36.52 billion barrels (63.2%) and the potential giants have 2.20 billion barrels (3.8%) of reserves. Continued discoveries of giant fields in Mexico should increase the proportion of reserves in giants.

The data available on North American giant oil fields are generally reliable. The major data uncertainties result from the as-yet-to-be-determined potential of several recently discovered fields, primarily in Mexico, and the uncertain potential of enhanced recovery in most of the major fields.

SOURCES:

1. American Gas Association, American Petroleum Institute, and Canadian Petroleum Association, *Reserves of Crude Oil, Natural Gas Liquids, and Natural Gas in the United States and Canada as of December 31, 1976*, Washington, D.C., 1977 (also previous years).

2. J. D. Moody, J. W. Mooney, and J. Spivak, "Giant Oil Fields of North America," in M. T. Halbouty, ed., *Geology of Giant Petroleum Fields*, American Association of Petroleum Geologists Memoir 14, Tulsa, Okla., 1970, pp. 8-17.

Table A.2

GIANT OIL FIELDS AND RECOVERABLE OIL RESOURCES OF CANADA AS OF DECEMBER 31, 1975

(In millions of barrels)

Rank/Field	Discovery Year	1975 Production	Cumulative Production	Reserves	Total Recovery (With Range of Estimates)		
Giants							
68. *Pembina*	1953	35.5	836	974	1,810	(?)	(1,660-1,810)
131. *Swan Hills*	1957	43.2	387	613	1,000		(930-1,000)
151. *Redwater*	1948	35.5	576	229	805		(700-820)
217. *Judy Creek*	1959	24.5	228	287	515		(490-540)
220. *South Swan Hills*	1960	22.7	169	341	510		(390-510)
Combination Giants							
246. *Weyburn-Midale*	1953	10.4	312	178	490		
254. *Bonnie Glen*	1952	19.6	250	195	445		(445-465)
265. *Leduc-Woodbend*	1947	6.9	335	30	365		(365-600)
272. *South Kaybob*	1963	2.9	28	62	90		
Others		301.6	4,286	4,924	9,210		
Canada Total		502.8	7,407	7,833	15,240		

NOTES: More information is available about oil fields in Canada than those in any other country in the world. As a result, there is less uncertainty about total recovery from Canadian fields than those anywhere else. The estimates for the individual fields and for Canada as a whole are taken from two different and independent sources, the provincial authorities and the Canadian Petroleum Association, respectively. The estimates of the latter for total proved and probable reserves tend to be somewhat higher than those of the provincial authorities, suggesting that a small amount of reserve growth could still occur in the giant fields listed here.

Pembina, Swan Hills, Judy Creek, and *South Swan Hills* have been waterflooded for some years. *Redwater* now has water injection to augment a strong natural water drive. Total recovery in all these fields but *Pembina* ranges from 40% to 65% of the oil-in-place. All five, particularly *Pembina,* have enhanced recovery potential. *Weyburn-Midale* (including *Benson, Huntoon, Lougheed, Midale,* and *Weyburn*) could prove to be an oil giant (currently total recovery is only 28% of estimated oil-in-place). *Bonnie Glen* has more than 125 million barrels of recoverable ethane and NGL besides the crude oil indicated here. *Leduc-Woodbend* is a marginal combination giant with nearly 125 million barrels of ethane and NGL plus several hundred billion cubic feet of dry gas. *South Kaybob* merits giant status by having about 450 million barrels of recoverable ethane and NGL in over 2.5 Tcf of wet gas.

Rainbow is often considered to be a giant field, but this collection of 75 individual reefal accumulations spread over 14 townships is a field for administrative purposes only. *West Pembina,* a 1977 discovery, has been characterized as a possible giant. The data about it released in early 1978 suggests that it will be a *Rainbow*-like set of reefal accumulations. Although there are several other large fields in Canada (most notably *Mitsui, Nipisi,* and *Wizard Lake*), none are large enough to be potential giants, even with enhanced recovery. *Crossfield* and *Waterton* are combination gas giants, each containing large amounts of gas liquids.

No survey of Canadian oil resources would be complete without mention of the massive heavy oil deposits in Alberta. These include *Athabasca* (626 billion barrels of 6°-10° API oil-in-place), *Cold Lake* (165 billion barrels of 10°-14° oil-in-place), *Peace River* (75 billion barrels of 6°-10° oil-in-place), and *Wabasca* (86 billion barrels of 6°-10° oil-in-place). An estimated 25%-50% of the oil-in-place in *Cold Lake* and 25%-33% of the oil-in-place in *Athabasca, Peace River,* and *Wabasca* may be recoverable. However, fuel use and processing losses during recovery and upgrading are likely to use 25%-50% of the oil recovered.

SOURCES:

1. Alberta Energy Resources Conservation Board, *Alberta's Reserves of Crude Oil, Gas, Natural Gas Liquids, and Sulfur at 31 December 1976,* Calgary, 1977 (also previous years).
2. Alberta Society of Petroleum Geologists, *Oil Fields of Alberta,* Calgary, 1960 (and 1967 supplement).
3. Department of Energy, Mines, and Resources, *Oil Sands and Heavy Oils: The Prospects,* Report EP 77-2, Ottawa, 1977.
4. Department of Mineral Resources, Province of Saskatchewan, *Petroleum and Natural Gas Reservoir Annual, 1975,* Regina, 1976.

Table A.3

GIANT OIL FIELDS AND RECOVERABLE OIL RESOURCES OF GUATEMALA AS OF DECEMBER 31, 1975

(In millions of barrels)

Rank/Field	Discovery Year	1975 Production	Cumulative Production	Reserves	Total Recovery (With Range of Estimates)
Others	--	--	--	25	25
Guatemala Total	--	--	--	25	25

NOTE: Oil was not discovered in Guatemala until 1975 in what appears to be a field of only modest dimensions.

Table A.4

GIANT OIL FIELDS AND RECOVERABLE OIL RESOURCES OF MEXICO AS OF DECEMBER 31, 1975

(In millions of barrels)

Rank/Field	Discovery Year	1975 Production	Cumulative Production	Reserves	Total Recovery (With Range of Estimates)		
Giants							
34. *Samaria-Cunduacan*	1973	59.5	79	4,421	4,500	(?)	
63. *Poza Rica*	1930	19.9	1,093	907	2,000		(2,000-2,750)
66. *Cactus-Nispero*	1972	26.3	46	1,804	1,850	(?)	
98. *Sitio Grande*	1972	26.8	43	1,207	1,250	(?)	
108. *Naranjos-Cerro Azul*	1909	1.6	1,155	20	1,175	(?)	(1,175-1,400)
135. *Ebano-Panuco*	1901	2.7	932	43	975		(970-1,300)
235. *Rio Nuevo*	1975	0.3	--	500	500	(?)	
Combination Giants							
248. *San Andres*	1956	11.1	259	221	480	(?)	
Possible Giants							
Cinco Presidentes	1960	7.3	200	60	260		
Others		106.0	1,766	1,744	3,510		
Mexico Total		261.5	5,573	10,927	16,500	(?)	(9,000-15,000)
Post-1975 Discoveries							
Agave	1976	--	--				(600-800)
Artesa	1977	--	--				
Mundo Nuevo	1976	--	--				
Zapatero	1977	--	--				

NOTES: Given the recent pace of discoveries in the Reforma area of Mexico, any listing of known and potential giant oil fields in Mexico is subject to revision almost every month. As a result, the estimates shown here for Mexican reserves is a highly conservative one. By the end of 1977, proved reserves in Mexico probably are double the estimate as of the end of 1975.

No estimates of reserves for the individual fields discovered in the Reforma area have been published. However, rough estimates can be inferred from the published estimates of total reserves for the producing fields in the area, published plans for production and water injection for each field, and descriptions of field characteristics. Such estimates are shown here for *Samaria-Cunduacan* (including *Crisol, Iride, Tierra Colorado,* and *Tres Pueblos*), *Cactus-Nispero, Sitio Grande,* and *Rio Nuevo.* These estimates may be conservative. Large water injection projects are planned to begin in all of these fields in the early years of their producing lives. They are expected to increase recovery from 20% (primary) to 45% of the oil-in-place. The evidence available was insufficient to determine whether the reserve estimates assumed primary recovery only or primary and secondary recovery, although some of the available reservoir volume parameters suggest that they are most likely for primary recovery only. Given the potential for reserve growth from both additional recovery and from extensions and new-pool discoveries, *Samaria-Cunduacan* could prove to be the first super-giant discovered in the 1970s.

Poza Rica and *San Andres* have substantial secondary recovery projects currently under way. One is scheduled to begin in *Cinco Presidentes* (which may prove to be only a combination giant). *Naranjos-Cerro Azul* and *Ebano-Panuco* are nearly depleted. (In each case, published reserve estimates for these fields were checked against Pemex estimates of proved reserves in the areas in which they are located.)

Of the post-1975 discoveries, both *Agave* and *Mundo Nuevo* are likely small giants (500-1500 million barrels of oil or gas in oil-equivalents). Several of the other 1976-1977 discoveries are likely to become recognized as giants as well. *Atun* and *Arenque,* two offshore discoveries of the 1960s, are erstwhile giants, with likely recoveries being only a fraction of initial estimates.

SOURCE: *Report of the Director General, Petroleos Mexicanos* (Pemex), issued annually on March 18.

Table A.5

GIANT OIL FIELDS AND RECOVERABLE OIL RESOURCES OF THE UNITED STATES AS OF DECEMBER 31, 1975
(In millions of barrels)

Rank/Field	Discovery Year	1975 Production	Cumulative Production	Reserves	Total Recovery (With Range of Estimates)		
Giants							
18. *Prudhoe Bay*	1968	0.7	2	9,598	9,600	(low?)	(8,800-20,000)
31. *East Texas*	1930	69.0	4,310	1,290	5,600		(5,100-6,000)
46. *Wilmington Trend**	1922	70.9	1,973	827	2,800	(low?)	(2,700-2,800)
56. *Yates*	1926	18.7	661	1,389	2,050	(low?)	(1,500-2,400)
67. *Midway-Sunset*	1901	37.3	1,265	565	1,830	(low?)	(1,200-2,000)
72. *Wasson*	1937	94.8	992	808	1,800	(low?)	(650-1,550)
74. *Scurry*	1948	79.1	1,051	699	1,750	(low?)	(1,190-1,750)
77. *Kern River*	1899	27.8	659	941	1,600	(low?)	(610-1,720)
85. *Slaughter-Levelland*	1936	60.6	885	615	1,500	(low?)	(690-1,500)
89. *Sho-Vel-Tum*	1915	32.8	1,034	391	1,425		(900-1,425)
90. *Panhandle*	1910	11.5	1,295	115	1,410		(1,410-1,645)
91. *Elk Hills*	1919	0.8	284	1,016	1,300	(low?)	(1,100-1,335)
106. *Ventura-Rincon**	1916	15.7	990	210	1,200	(low?)	(1,000-1,200)
113. *Huntington Beach**	1920	17.2	956	139	1,095	(low?)	(930-1,095)
122. *Hawkins*	1940	40.8	577	438	1,015		(525-1,070)
137. *Greta-Tom O'Connor*	1933	29.2	643	307	950		(570-950)
140. *Long Beach*	1921	2.5	871	59	930	(low?)	(880-930)
141. *Eunice Area*	1927	12.7	655	245	900	(low?)	
148. *Goldsmith-Andector*	1935	20.3	674	181	855	(low?)	(610-860)
150. *Coalinga*	1887	6.0	639	191	830	(low?)	(630-830)
155. *McElroy-Dune*	1926	19.3	474	326	800	(low?)	(490-915)
160. *Rangely*	1902	20.5	534	261	795	(low?)	(535-795)
161. *Hastings*	1935	28.1	531	234	765		(500-765)
163. *Salt Creek*	1889	10.0	560	200	760		(560-760)
164. *Oklahoma City*	1928	1.9	736	19	755	(low?)	(750-770)
168. *Caillou Island**	1930	14.1	529	206	735	(?)	(500-735)
169. *Conroe*	1931	21.4	558	177	735	(low?)	(675-735)
180. *Bradford*	1871	2.1	661	19	680	(low?)	(655-680)
181. *Buena Vista*	1910	3.4	619	31	650	(low?)	(610-650)
183. *Webster*	1936	25.1	439	201	640		(340-640)
185. *Santa Fe Springs*	1919	0.7	602	23	625	(low?)	(610-625)
186. *Bay Marchand Block 2**	1949	28.3	461	149	610	(low?)	(575-650)
203. *Spraberry Trend*	1952	18.2	418	152	570	(low?)	(280-575)
209. *Van*	1928	15.8	435	115	550	(low?)	(405-550)
210. *Burbank*	1920	3.4	507	33	540	(low?)	(500-540)

Table A.5--continued

Rank/Field	Discovery Year	1975 Production	Cumulative Production	Reserves	Total Recovery (With Range of Estimates		
211. *Smackover*	1922	3.1	512	28	540	(low?)	(525-530)
212. *Elk Basin*	1915	8.0	449	86	535	(low?)	(400-545)
213. *San Ardo*	1947	13.9	288	242	530	(low?)	(275-660)
214. *McArthur River***	1965	40.9	294	231	525		(160-525)
218. *South Sand Belt*	1926	3.7	493	22	515	(low?)	
219. *Borregos-Seeligson-T.C.B.*	1938	4.8	478	32	510		(575-625)
225. *Coalinga East Extension*	1938	3.3	472	28	500		(495-500)
241. *Thompson (all)*	1921	15.2	384	116	500		(350-500)
Combination Giants							
244. *Cushing*	1912	2.7	466	24	490	(low?)	(455-485)
247. *Timbalier Bay**	1939	11.6	410	75	485		(475-685)
249. *South Cowden, Foster, Johnson*	1932	16.2	331	139	470	(low?)	(320-470)
250. *North Cowden*	1930	16.5	298	167	465	(low?)	(260-465)
251. *South Pass Block 24**	1950	12.8	382	83	465		(455-750)
252. *Kettleman North Dome*	1928	0.5	453	7	460	(low?)	(455-475)
255. *Brea-Olinda-Sansinena*	1884	3.9	395	45	440	(low?)	(405-440)
256. *Vacuum*	1929	14.0	292	148	440	(low?)	(285-440)
257. *Seminole*	1936	21.7	211	219	430	(low?)	(200-430)
258. *West Delta Block 30***	1954	17.7	332	93	425	(low?)	(300-450)
260. *Fullerton*	1941	7.3	260	155	415	(low?)	(275-415)
263. *South Pass Block 27***	1954	9.7	268	107	375		(220-385)
264. *West Ranch*	1938	13.8	295	80	375		(280-425)
268. *Grand Isle Block 43***	1956	17.6	181	94	275		(275-370)
269. *Blinebry-Drinkard*	1944	4.4	199	51	250		
271. *Sand Hills*	1930	4.6	190	45	235	(low?)	(200-245)
Probable Giants							
South Belridge	1911	9.3	206	214	420	(low?)	(200-420)
Caddo-Pine Island	1906	3.1	320	25	345	(low?)	(340-390)
*Eugene Island Block 330***	1971	31.1	62	238	300	(low?)	(150-230)
Healdton	1912	6.7	301	49	350	(low?)	(270-350)
*Hondo***	1969	--	--	95	95	(low?)	(95 and up)
Howard Glasscock	1925	7.0	317	113	430	(low?)	(285-430)
Loudon	1938	2.3	361	24	385	(low?)	(365-385)
Oregon Basin	1912	12.6	253	122	375	(low?)	(185-505)

Table A.5--continued

Rank/Field	Discovery Year	1975 Production	Cumulative Production	Reserves	Total Recovery (With Range of Estimates)		
Possible Giants							
Altamont-Bluebell	1970	23.2	72	88	160	(?)	(150-1,000)
Anahuac	1935	7.3	251	49	300		(265-355)
Aneth	1956	8.3	269	81	350	(low?)	(315-450)
Artesia-Maljamar	1923	8.6	272	68	340	(low?)	
Cat Canyon	1908	6.4	222	78	300	(low?)	(250-300)
Hobbs	1928	4.5	232	128	360		(215-360)
Jay	1970	33.8	111	234	345	(low?)	(250-345)
Jordan-Waddell-Penwell-Edwards	1927	7.3	283	62	345	(low?)	
Kern Front	1915	4.0	139	61	200	(low?)	(150-200)
Keystone	1930	3.7	273	27	300		(300-320)
Lake Washington	1931	5.3	202	58	260		(260-300)
Lawrence	1906	3.2	371	29	400	(low?)	(365-400)
*Main Pass Block 41***	1957	8.9	152	68	220		(220-280)
McKittrick: Main	1887	4.0	163	37	200	(low?)	(160-200)
Mount Poso	1926	3.5	173	47	220	(low?)	(175-220)
Pegasus	1949	2.3	115	15	130	(low?)	
*Point Thompson Unit**	1975	--	--				
Salem	1939	2.8	361	29	390	(low?)	(365-375)
Santa Maria Valley	1934	3.0	162	63	225	(low?)	(160-225)
North Ward-Estes	1929	5.4	334	36	370	(low?)	(305-375)
Others		1,540.0	66,219	12,366	78,385		
United States Total		2,886.3	109,008	38,992	148,000		
Post-1975 Discoveries							
*San Pedro OCS Blocks 254, 261, and 262***	1976	--	--				

NOTES: There are more known and potential giant oil fields in the United States than in any other country. Most, however, are small giant fields with ultimate recovery of less than 1 billion barrels. Although most are quite old (only 13 of the 87 known and potential giants were discovered after 1950), estimates of total recovery in many of these fields are still uncertain, primarily because enhanced recovery operations have only just begun or are still in the planning stages. As a result, there is still considerable room for reserve growth in many of these fields. The estimates of current reserves used here are principally those of the American Petroleum Institute, amended in those cases where production trends and operators' plans indicate likely understatement. They include both proved reserves and indicated additions to reserves.

The estimate for *Prudhoe Bay* (Alaska) is for the Sadlerochit sandstone only, excluding heavier oil discoveries in the Kuparuk sandstone and Lisburne formation. The former may be a separate field or several fields; the latter is likely to be included in the *Prudhoe Bay* field, increasing its total reserves. The effect of secondary recovery in the field is still undetermined. The highly efficient water flood in *East Texas* (Texas) is resulting in 75%-80% recovery of the estimated oil-in-place. Higher prices could increase recovery in the *Wilmington Trend* (including *Torrance* and *Belmont*) (California) by several hundred million barrels. Further increases in allowable production in *Yates* (including *Toborg*) (Texas) together with the use of carbon dioxide miscible flooding could increase recovery by another half billion barrels or more. The continued application of thermal recovery in the heavy oil *Midway-Sunset* and *Kern River* (both California) fields could increase total recovery from the amounts shown here by 50% or more. Infill drilling and the use of carbon dioxide miscible flooding is expected to increase total recovery in *Wasson*, *Scurry* (*Kelly-Snyder-Diamond M*), and *Slaughter-Levelland* (all Texas) by several hundred million barrels apiece. Production in *Sho-Vel-Tum* (Oklahoma) and *Panhandle* (Texas) is slowly declining. New-pool discoveries and the application of secondary recovery in *Elk Hills* (California) are likely to increase reserves there. Production in *Ventura-Rincon* and *Huntington Beach* (California) is declining, but there is still some potential for small reserve increases. With total recovery in *Hawkins* (Texas) already at an estimated 70%-75% of oil-in-place, it has little room for reserve growth.

Table A.5--continued

Because current recovery in *Greta-Tom O'Connor* (Texas) is already high (60%), its potential for reserve growth is small. The recently initiated water flood in *Long Beach* (California) is likely to increase recovery by even more than the reserves shown here. Use of carbon dioxide miscible flooding in the carbonate reservoirs of *Eunice Area* (including *Eumont, South Eunice, Eunice-Monument, Jalmat, Langlie-Mattix, Monument, Fowler,* and *Justis*) (New Mexico), *Goldsmith-Andector* (Texas), and *McElroy-Dune* (Texas) could produce 20%-50% increases in total recovery. Expanded use of thermal recovery methods is expected to increase recovery in *Coalinga* (California). The extensive water flood in *Rangely* (Colorado) is resulting in a 45% recovery factor. After being on a plateau for several years, production in *Hastings* (Texas) is on the verge of a steady decline. Several recently initiated water floods in *Salt Creek* (including *Teapot Dome*) (Wyoming) have increased its reserves. *Oklahoma City* (Oklahoma) and *Buena Vista* (California) appear to be largely depleted, but both have some enhanced recovery potential. Production in *Caillou Island* (Louisiana) is declining, but there are several pools in the field that are currently not producing. The proposed unitization plan for *Conroe* (Texas) is expected to increase recovery by more than 50 million barrels beyond the level shown here, raising total recovery to nearly 65%. *Bradford* (Pennsylvania), the oldest U.S. giant, continues to produce.

Webster (Texas) has undergone both gas injection and water flooding and is unlikely to see any major reserve increases in the future. Expansion of the water flood in *Santa Fe Springs* (California) should increase recovery there. Production in *Bay Marchand Block 2* (Louisiana) is declining sharply, but some reserve growth is possible if a carbon dioxide miscible flood could be implemented in the field. Because the tight reservoirs of the *Spraberry Trend* (Texas) have the lowest recovery factor of any giant field in the United States (less than 7%), they provide a large target for additional recovery efforts. Production in *Van* (Texas) should begin to decline steadily in 1977 or 1978, a decline that could be slowed by enhanced recovery. Production in *Burbank* (Oklahoma) has been declining, but total recovery could be increased by at least 15%-25% from the application of enhanced recovery methods. The application of thermal recovery methods in *Smackover* (Arkansas) and *San Ardo* (California) is increasing total recovery in both fields. Although recovery is already over 50%, *Elk Basin* (Wyoming and Montana) has substantial enhanced recovery potential. The *South Sand Belt* (the overlapping and tangential *Emperor, Halley, Henderson, Hendrick, Kermit, Scarborough,* and *Weiner* fields) (Texas and New Mexico) is largely depleted, though its carbonate reservoirs have some enhanced recovery potential. *Borregos-Seeligson-T.C.B.* (Texas) is also a giant gas field. As there has been little secondary recovery in *Coalinga East Extension* (California), it could still have some potential for reserve growth. Production in *Thompson* has begun a steady and probably irreversible decline.

Because a sizeable market for natural gas has existed in the United States, the necessary data are available to identify combination giant fields. *Cushing* (Oklahoma), *South Cowden, Foster, Johnson* (Texas), and *North Cowden* (Texas) could easily become full oil giants with some additional recovery. Production in *Timbalier Bay* and *South Pass Block 24* (both Louisiana) has been declining, but both have enhanced recovery possibilities. *Kettleman North Dome* (California), with 2.97 Tcf of gas, narrowly misses being both a giant oil field and a giant gas field. *Brea-Olinda-Sansinena* (California) is a marginal combination giant. If carbon dioxide miscible flooding is applied to *Vacuum* (New Mexico) and *Seminole* (Texas), both will easily become full oil giants. Production has been declining sharply in *West Delta Block 30* and *South Pass Block 27* (both Louisiana), but carbon dioxide miscible flooding may be applied in each. *Fullerton* (Texas) also has enhanced recovery possibilities. Oil production in *West Ranch* (Texas) is declining and the possibilities of enhanced recovery are not promising. Production in *Grande Isle Block 43* (Louisiana) is declining sharply. Both *Blinebry-Drinkard* (including *Penrose-Skelly, Paddock, Tubb, Hare,* and *Brunson*) (New Mexico) and *Sand Hills* (Texas) have had substantial natural gas liquids as well as crude oil production.

Because of the application of thermal recovery, production in *South Belridge* (California) is increasing, making it likely that it will be recognized as a giant within a few years. With surfactant flooding, *Caddo-Pine Island* (Louisiana), *Healdton* (Oklahoma), and *Loudon* (Illinois) are expected to become giants. Production in *Eugene Island Block 330* (Louisiana) reached 90 Mb/d and 200 MMcfd by late 1977, and the field is still not fully developed. It is a likely combination giant. The estimates for *Hondo* (California) are for recovery from the initial platform only. It is the largest of three fields in the *Santa Ynez Unit*, which has 800 to 1100 million barrels of reserves. *Howard Glasscock* (Texas) is expected to become a giant from the application of carbon dioxide miscible flooding. The current water flood in *Oregon Basin* (Wyoming) is predicted to make it a marginal giant.

Altamont-Bluebell (Utah) may become a giant if higher prices permit extensive infill drilling (at 80 to 160 acre spacing instead of the present 640 acre spacing). With enhanced recovery, *Anahuac* (Texas) could become a marginal combination giant. *Aneth* (Utah) is likely to become a giant with the application of carbon dioxide miscible flooding. The application of enhanced recovery could make *Artesia-Maljamar* and *Hobbs* (both New Mexico) combination giants. *Cat Canyon, Kern Front, McKittrick: Main, Mount Poso,* and *Santa Maria Valley* (all California) are predicted to become giants with the application of thermal recovery. If the water flood in *Jay* (Florida) continues to exceed expectations, it could become a marginal combination giant. There has been extensive water flooding in *Jordan-Waddell-Penwell-Edwards* and *Keystone* (both Texas), but with enhanced recovery they could become combination giants. If carbon dioxide miscible flooding is applied in *Lake Washington* and *Main Pass Block 41* (both Louisiana), both could become combination giants. *Lawrence* and *Salem* (both Illinois) may become giants if surfactant flooding can be applied successfully in each. *Pegasus* (Texas) might prove to be a marginal combination giant once carbon dioxide miscible flooding is applied. The *Point Thompson Unit* (Alaska) is a heavy oil (18°-23° API) field of clearly giant dimensions. *North Ward-Estes* (Texas) is another likely combination giant with enhanced recovery. The structure in *San Pedro OCS Blocks 254, 261,* and *262* (California) clearly has the reservoir volume to be a giant, but the deep-water (200-800 feet) location of the field, its heavy oil (14°-20° API), and unconsolidated sandstone reservoirs may preclude recovery in significant amounts.

Table A.5--continued

There are several other fields or groups of fields in the United States that have been included in other compilations of giant fields. *West Delta Block 73*** (Louisiana) is excluded here because it has proved to be much smaller than thought initially. *Golden Trend* (Oklahoma), *Sooner Trend* (Oklahoma), *Seminole Plateau* (Oklahoma), *Lima-Indiana* (Ohio-Indiana), and *Wichita County Regular* (Texas) are complexes of several laterally distinct pools, none of which by itself is of giant status. All are excluded here because they do not merit single field status by the criteria used in this report.

SOURCES:

1. Alaska Department of Natural Resources, Division of Oil and Gas Conservation, *Statistical Report for the Year 1976*, Anchorage, 1977 (also previous years).

2. Arkansas Oil and Gas Commission, *Annual Oil and Gas Report, 1975*, El Dorado, 1976 (also previous years).

3. California Division of Oil and Gas, *Annual Report of the State Oil and Gas Supervisor 1976*, Sacramento, 1977 (also previous years).

4. Colorado Oil and Gas Conservation Commission, *Oil and Gas Statistics 1975*, Denver, 1976.

5. Federal Energy Administration, *Oil and Gas Resources, Reserves, and Productive Capacities*, Vol. II, Washington, D.C., October 1975.

6. M. T. Halbouty, "Giant Oil and Gas Fields in United States," in M. T. Halbouty, ed., *Geology of Giant Petroleum Fields*, American Association of Petroleum Geologists Memoir 14, Tulsa, Okla., 1970, pp. 91-127.

7. Illinois State Geological Survey, *Petroleum Industry in Illinois, 1975*, Urbana, 1976.

8. K. K. Landes, *Petroleum Geology of the United States*, John Wiley and Sons, New York, 1970.

9. Louisiana Department of Conservation, *Annual Oil and Gas Report 1974*, Baton Rouge, 1977 (also previous years).

10. Montana Oil and Gas Conservation Division, *Annual Review for the Year 1976 Relating to Oil and Gas*, Helena, 1977 (also previous years).

11. National Petroleum Council, *Enhanced Oil Recovery*, Washington, D.C., December 1976.

12. R. Nehring, *The Discovery of Significant Oil and Gas Fields in the United States*, The Rand Corporation, Santa Monica, Calif., forthcoming 1979.

13. New Mexico Oil and Gas Engineering Committee, *Annual Report 1976*, Hobbs, 1977 (also previous years).

14. Office of Technology Assessment, *Enhanced Oil Recovery Potential in the United States*, Washington, D.C., January 1978.

15. Railroad Commission of Texas, Oil and Gas Division, *Oil and Gas Annual Production by Active Fields, 1975*, Austin, 1976.

16. Railroad Commission of Texas, Oil and Gas Division, *A Survey of Secondary and Enhanced Recovery Operations in Texas to 1976*, Bulletin 76, Austin, 1977.

17. Wyoming Oil and Gas Conservation Commission, *Wyoming Oil and Gas Statistics, 1976*, Casper, 1977 (also previous years).

Table A.6

KNOWN RECOVERABLE CRUDE OIL RESOURCES OF SOUTH AMERICA BY FIELD CLASSIFICATION AS OF DECEMBER 31, 1975
(In millions of barrels)

Field Classification	Number of Fields	1975 Production	Cumulative Production	Reserves	Total Recovery
Giant fields	19	802.7	28,905	17,800	46,705
Combination giants	--	--	--	--	--
Probable giants	2	13.0	697	203	900
Possible giants[a]	11	71.6(+)	1,386(+)	2,099(+)	3,485(+)
Other fields[a]		419.9(-)	9,620(-)	7,710(-)	17,330(-)
Total		1,307.2	40,608	27,812	68,420

[a]Because no data could be obtained on *Tupungata*, a potential giant field in Argentina, its production and reserves are included with those of the non-giant fields. As a result, the estimates for possible giants are understated and those for other fields are overstated.

NOTES: South America is unique among the oil-producing regions of the world in that nearly half of the total known recoverable crude oil resources of the continent are in one field, *Bolivar Coastal* of Venezuela. Despite this unusual concentration, giant oil fields contain slightly less of the known recoverable crude oil in South America than they do worldwide. The 19 known giant fields contain 46.71 billion barrels (68.3%), and the 13 potential giants contain somewhat more than 4.39 billion barrels (6.4%). Of the 17.33 billion barrels in other fields, over 8.5 billion are in at least 53 large fields.

Giant oil fields appear to be slightly less important for future production than for past production in South America. Known giant fields contain only 64.0% of reserves and potential giants only 8.3%. However, this may be only a statistical artifact, as most of the potential for future reserve growth is in the giant fields.

Except for Argentina, reasonably reliable data on past and current production are readily available for all major fields in South America. Estimates of reserves are of mixed quality; but overall the margins of possible error do not appear to be overly large.

SOURCE: L. E. Hatfield and C. H. Neff, "Petroleum Developments in South America, Central America, and Caribbean Area in 1975," *American Association of Petroleum Geologists Bulletin*, Vol. 60, No. 10, October 1976, pp. 1640-1703.

Table A.7

GIANT OIL FIELDS AND RECOVERABLE OIL RESOURCES OF ARGENTINA AS OF DECEMBER 31, 1975
(In millions of barrels)

Rank/Field	Discovery Year	1975 Production	Cumulative Production	Reserves	Total Recovery (With Range of Estimates)	
Possible Giants						
Tupungata	1934		*(Information not available)*			
Others (includes possible)		144.1	2,640	2,560	5,200	
Argentina Total		144.1	2,640	2,560	5,200	(5,100-5,400)

NOTES: Recent information about production in Argentina is provided by province only. *Tupungata,* in Mendoza province, is characterized by Tiratsoo as a major factor in the growth of the oil industry in Argentina. All the other fields in the province are stated to be small. Given an estimated total recovery of 1 billion barrels in the province (or more with extensive secondary recovery), *Tupungata* is a likely giant. *Comodoro Rivadavia* is often included as a giant field. It is more accurately described as a producing area containing several fields. The published estimates of reserves in Argentina indicate only part of the existing potential for secondary recovery there, the amount for total recovery shown here being less than 20% of oil-in-place.

Table A.8

GIANT OIL FIELDS AND RECOVERABLE OIL RESOURCES OF BOLIVIA AS OF DECEMBER 31, 1975
(In millions of barrels)

Rank/Field	Discovery Year	1975 Production	Cumulative Production	Reserves	Total Recovery (With Range of Estimates)	
Others		14.7	187	233	420	
Bolivia Total		14.7	187	233	420	(325-425)

NOTES: Most of Bolivian production comes from six fields with an estimated total recovery of 10 to 100 million barrels each. Discoveries made during 1976 nearly doubled reserves.

Table A.9

GIANT OIL FIELDS AND RECOVERABLE OIL RESOURCES OF BRAZIL AS OF DECEMBER 31, 1975

(In millions of barrels)

Rank/Field	Discovery Year	1975 Production	Cumulative Production	Reserves	Total Recovery (With Range of Estimates)
Possible Giants					
*Namorado***	1975	0	0	440	440 (?) (400-440)
Others		62.8	855	1,455	2,310
Brazil Total		62.8	855	1,895	2,750 (?) (1,625-2,200)

NOTES: Despite an extensive and expensive exploration effort over the past quarter century in Brazil, no known giants have yet been discovered. Further extensions to *Namorado* may prove it to be the first. However, the field will be expensive to develop given its deep-water (500-600 feet) location 50 miles from shore. *Garoupa,* immediately to the north of *Namorado* and perhaps in the same mosaic of fault blocks, was heavily touted as a giant after its discovery in 1974 before being subsequently downgraded following further drilling.

Most of Brazilian production to date has come from five fields with estimated total recovery between 100 and 400 million barrels (*Agua Grande, Aracas, Buracica, Carmopolis,* and *Miranga*). *Agua Grande,* the largest, will approach giant status with secondary recovery and its oil-equivalent gas production, but its likely maximum total recovery will be between 400 and 450 million barrels of oil and oil-equivalent gas.

Besides the Campos Basin fields (*Namorado, Garoupa,* and others), two other large but non-giant fields (*Agulha* and *Uburana*) have been discovered in Brazil in the past five years. The estimated reserves from all these new discoveries (roughly 1150 million barrels) are included in the estimate of total Brazilian reserves.

Table A.10

GIANT OIL FIELDS AND RECOVERABLE OIL RESOURCES OF CHILE AS OF DECEMBER 31, 1975

(In millions of barrels)

Rank/Field	Discovery Year	1975 Production	Cumulative Production	Reserves	Total Recovery (With Range of Estimates)
Others		9.2	216	184	400
Chile Total		9.2	216	184	400 (400-425)

NOTES: Recent discoveries in Chile doubled reserves between 1975 and 1977. None approach the giant category.

Table A.11

GIANT OIL FIELDS AND RECOVERABLE OIL RESOURCES OF COLOMBIA AS OF DECEMBER 31, 1975

(In millions of barrels)

Rank/Field	Discovery Year	1975 Production	Cumulative Production	Reserves	Total Recovery (With Range of Estimates)	
Giants						
175. *Infantas-La Cira*	1918	7.4	626	74	700	
Others		49.3	1,316	534	1,850	
Colombia Total		56.7	1,942	608	2,550	(2,500-2,650)

NOTES: *Infantas-La Cira,* one of the earliest discoveries in Colombia, is also the largest discovery to date. *Orito* has been asserted by one source to be a giant, but given its production history and estimates of total Colombian reserves from the same source, it is clearly not one. Roughly 70% of the remainder of Colombian production comes from seven fields with an estimated total recovery between 100 and 300 million barrels each (*Casabe, Orito, Provincia, Rio Zulia, Tibú, Velásquez,* and *Yariqui*). All were discovered between 1940 and 1965.

Table A.12

GIANT OIL FIELDS AND RECOVERABLE OIL RESOURCES OF ECUADOR AS OF DECEMBER 31, 1975

(In millions of barrels)

Rank/Field	Discovery Year	1975 Production	Cumulative Production	Reserves	Total Recovery (With Range of Estimates)		
Giants							
158. *Shushufindi*	1969	26.6	87	713	800	(?)	(410-850)
197. *Sacha*	1969	16.8	86	514	600	(?)	(440-1,400)
Possible Giants							
Auca	1970	2.6	3	297	300	(?)	(to 500)
Others		12.7	158	242	400		
Ecuador Total		58.7	334	1,766	2,100	(?)	(1,700-2,800)

NOTES: Estimates of the potential of the fields discovered in the Oriente region between 1967 and 1970 have fluctuated considerably, resulting in similar fluctuations in estimates of total Ecuadorian reserves. During this period production has been constrained by government production restrictions, pipeline breaks, a lack of storage capacity, and a lack of demand. Further development has been constrained by disagreements between the government and the operators of the field. These constraints appear to be the primary reasons for the uncertainty in the estimates.

Given favorable economic conditions, both *Shushufindi* and *Sacha* are clearly giants. With successful secondary recovery, both could be somewhat larger. *Auca* may be a giant, provided successful secondary recovery occurs. The only other large field that is producing now is *Lago Agrio,* with an estimated 250 million barrels' total recovery. There are several other discoveries in the Oriente region that have not yet been developed and are not included in the reserve estimate used here. They could add as much as 800 million barrels to total reserves if they are developed.

Table A.13

GIANT OIL FIELDS AND RECOVERABLE OIL RESOURCES OF PERU AS OF DECEMBER 31, 1975

(In millions of barrels)

Rank/Field	Discovery Year	1975 Production	Cumulative Production	Reserves	Total Recovery (With Range of Estimates)
Giants					
205. *La Brea-Parinas*	1869	3.4	510	50	560 (?) (to 1,000)
Others		23.0	431	709	1,140 (?)
Peru Total		26.4	941	759	1,700 (?) (1,600-1,900)

NOTES: *La Brea-Parinas* is one of the first giant fields to be discovered worldwide. It still has the potential for some reserve growth from new pool discoveries and additional recovery, although it appears most unlikely to reach the billion-barrel mark. Since 1970, there has been considerable exploration in the Oriente region of Peru resulting in several small to medium discoveries. The known results of these discoveries are included in the estimates of total reserves for Peru.

Table A.14

GIANT OIL FIELDS AND RECOVERABLE OIL RESOURCES OF TRINIDAD AND TOBAGO AS OF DECEMBER 31, 1975

(In millions of barrels)

Rank/Field	Discovery Year	1975 Production	Cumulative Production	Reserves	Total Recovery (With Range of Estimates)
Giants					
167. *Fyzabad-Forest Reserve*	1913	8.7	660	90	750 (?)
216. *Soldado***	1954	17.6	281	239	520 (?)
Others		52.3	612	618	1,230
Trinidad and Tobago Total		78.6	1,553	947	2,500 (?) (2,200-2,550)

NOTES: *Fyzabad-Forest Reserve* (considered here to consist of all the overlapping and tangential pools of Areas 4, 5, and 6, including *Palo Seco* and *Coora-Quarry*) contains mostly heavy oil (20°-30° API) and thus has the potential for some reserve growth with enhanced recovery. *Soldado* is another heavy oil field (25° API) which may possibly prove to be somewhat larger. Most of the remainder of current production and reserves in Trinidad is in three recently discovered offshore fields, *Poui*, *Samaan*, and *Teak*. Each appears to be about 200-300 million barrels in size.

SOURCE: K. W. Barr, S. T. Waite, and C. C. Wilson, "The Mode of Oil Occurrence in the Miocene of Southern Trinidad, B.W.I.," in L. G. Weeks, ed., *Habitat of Oil*, American Association of Petroleum Geologists, Tulsa, Okla., 1958, pp. 533-550.

Table A.15

GIANT OIL FIELDS AND RECOVERABLE OIL RESOURCES OF VENEZUELA AS OF DECEMBER 31, 1975

(In millions of barrels)

Rank/Field	Discovery Year	1975 Production	Cumulative Production	Reserves	Total Recovery (With Range of Estimates)
Giants					
3. *Bolivar Coastal**	1917	566.0[E]	20,200[E]	11,800	32,000 (low?) (16,500-35,000)
75. *Lamar***	1958	44.0[E]	745[E]	905	1,650 (low?) (1,000-1,410)
102. *Centro***	1957	40.0[E]	406[E]	794	1,200 (low?) (200-725)
118. *Boscan*	1946	18.0[E]	516[E]	524	1,040 (low?) (675-1,040)
132. *Urdaneta***	1955	--	2	998	1,000 (low?) (350-2,000)
138. *La Paz*	1923	6.5	791	159	950 (?) (905-1,600)
145. *Oficina*	1937	10.6	664	236	900 (?) (610-1,150)
146. *Quiriquire*	1928	4.9	733	167	900 (low?) (810-1,500)
173. *Guara-Guico*	1944	9.0[E]	588	112	700 (550-900)
177. *Mene Grande*	1914	4.9	587	113	700 (low?) (600-1,200)
196. *Mata*	1954	9.6[E]	456	144	600 (?) (230-1,100)
201. *Nipa*	1945	6.9	442	133	575 (?) (340-720)
204. *Greater Jusepin*	1938	1.8[E]	525[E]	35	560 (low?)
Probable Giants					
Mara	1945	4.0	383	67	450 (low?) (425-750)
Santa Rosa	1955	9.0[E]	314[E]	136	450 (low?) (240-470)
Possible Giants					
Chimire	1948	3.4	345	55	400 (?) (315-625)
Dacion	1957	4.3	178	92	270 (low?) (200-270)
Jobo	1956	15.6	91	309	400 (low?)
*Lago***	1958	18.3	141	259	400 (?)
Merey	1937	7.0[E]	185[E]	165	350 (low?) (190-250)
Morichal	1958	6.7	122	178	300 (low?) (190-250)
Oritupano	1950	7.0[E]	112[E]	188	300 (low?) (100-290)
Sinco	1953	6.7	209	116	325 (low?) (215-360)
Others		51.8	3,205	1,175	4,380
Venezuela Total		856.0	31,940	18,860	50,800 (?) (48,000-50,500)

NOTES: Recently published estimates of current and cumulative production in the major oil fields of Venezuela are slightly shaky, with minor differences among the various sources and frequent inconsistencies in time series from the same source. Reserve estimates are also uncertain, primarily because many of the fields are heavy oil fields whose reserves can be increased substantially through the implementation of thermal recovery. There is also some potential for additional secondary recovery in the fields with light- and medium gravity crude oil.

The *Bolivar Coastal* field (consisting of *Bachaquero*, *Cabimas*, *Ceuta*, *Lagunillas*, *Lama*, and *Tia Juana*) was the first super-giant oil field discovered in the world. Because primary recovery in many of its heavy oil reservoirs was quite low (15%-20%), the extensive application of thermal recovery could increase total recovery by 50% or more. (Production in the field has been curtailed substantially since 1973.) Total recovery in the two most recently discovered Venezuelan giants, *Lamar* and *Centro*, could increase because of secondary recovery. (Some sources consider *Centro* to be part of the *Bolivar Coastal* field.) *Boscan*, with its 10° API oil, is a clear candidate for the application of thermal recovery methods, although the depth of the reservoir may limit their feasibility. *Urdaneta*, another heavy oil field, has yet to be extensively developed, but at current prices it is clearly a giant. *La Paz* appears to be largely depleted. *Oficina*, with its 22° API oil, is a thermal recovery candidate. Thermal recovery operations are being initiated in *Quiriquire* (16° API) and have been occurring for several years in *Mene Grande* (18° API). *Guara-Guico* may have

Table A.15--continued

some secondary recovery potential. Both *Mata* and *Nipa* appear to be only marginal giants, as production in both has declined considerably. Thermal recovery has recently begun in *Greater Jusepin* (*Jusepin, Mata Grande, Mulata, Muri, Pirital, Santa Barbara,* and *Tacat)*, suggesting a potential for considerable reserve growth. *Mara* and *Santa Rosa* could already be combination giants. *Chimire* and *Sinco* could become giants with secondary or enhanced recovery. *Dacion, Jobo, Merey, Morichal,* and *Oritupano* are all heavy oil fields (10°-20° API) with the potential of becoming giants from the application of thermal recovery. *Lago* could become a giant from secondary recovery.

Large oil resources are also found in Venezuela in the *Orinoco* heavy oil belt. This area is estimated to contain hundreds of billions, if not trillions, of barrels of 4°-12° API oil-in-place. The oil also has a high sulfur (2%-5%) and a high metals content, making it expensive to process as well as to extract.

SOURCES:

1. A. R. Martinez, "Giant Fields of Venezuela," in M. T. Halbouty, ed., *Geology of Giant Petroleum Fields*, American Association of Petroleum Geologists Memoir 14, Tulsa, Okla., 1970, pp. 326-336.

2. J. B. Miller et al., "Habitat of Oil in the Maracaibo Basin, Venezuela," in L. G. Weeks, ed., *Habitat of Oil*, American Association of Petroleum Geologists, Tulsa, Okla., 1958, pp. 601-640.

3. H. H. Renz et al., "The Eastern Venezuelan Basin," in L. G. Weeks, ed., *Habitat of Oil*, American Association of Petroleum Geologists, Tulsa, Okla., 1958, pp. 551-600.

Table A.16

KNOWN RECOVERABLE CRUDE OIL RESOURCES OF WESTERN EUROPE BY FIELD CLASSIFICATION AS OF DECEMBER 31, 1975

(In millions of barrels)

Field Classification	Number of Fields	1975 Production	Cumulative Production	Reserves	Total Recovery
Giant fields	11	71.4	110	12,080	12,190
Combination giants	5	11.4	364	1,516	1,880
Probable giants	6	--	--	2,530	2,530
Possible giants	7	6.8	200	1,475(+)	1,675(+)
Other fields		110.8	2,330	4,040	6,370
Total		200.4	3,004	21,641(+)	24,645(+)

NOTES: Because of the numerous discoveries in the North Sea since the late 1960s, giant oil fields are now as important for the oil resources of Western Europe as they are for the rest of the world. The 16 known giant fields contain 14.07 billion barrels, 57.1% of the crude oil discovered and made recoverable in Western Europe to the end of 1975. The 13 potential giants contain at least 4.21 billion barrels, 17.1% of the total. Only 6.37 billion barrels (25.8%) are in non-giant fields. Of this amount, 3.65 billion barrels are found in 22 large fields.

Giant oil fields are even more important for future production in Western Europe. Known giants currently have 13.60 billion barrels of reserves, 62.8% of the total. Potential giants have at least 4.01 billion barrels, accounting for another 18.5%. The remaining fields have only 4.04 billion barrels of reserves, only 18.7% of the total.

As the oil fields of the North Sea are fully delineated, developed, and brought into production, the data on Western European giant oil fields are becoming more reliable. There are still some uncertainties to be resolved, such as the extent to which subsea production and injection systems will be used in order to recover a greater proportion of known in-place resources and the accuracy of predictions about producing lives of the known fields, but the answers to these should become apparent within the next few years.

SOURCES:

1. *European Petroleum Year Book, 1976*, Otto Vieth Verlag, Hamburg, 1977.
2. R. E. King, "Petroleum Exploration and Production in Europe in 1975," *American Association of Petroleum Geologists Bulletin*, Vol. 60, No. 10, October 1976, pp. 1704-1766 (also previous years).

Table A.17

GIANT OIL FIELDS AND RECOVERABLE OIL RESOURCES OF AUSTRIA AS OF DECEMBER 31, 1975

(In millions of barrels)

Rank/Field	Discovery Year	1975 Production	Cumulative Production	Reserves	Total Recovery (With Range of Estimates)	
Combination Giants						
245. *Greater Matzen*	1949	10.9	363	127	490	
Others		3.3	188	42	230	
Austria Total		14.2	551	169	720	(720)

NOTES: Oil production in Austria is dominated by *Greater Matzen* (including the deeper, partially underlying oil and gas fields of *Prottes-Tief, Reyersdorf, Schoenkirch, Strasshof-Tief,* and *Tallesbrunn*). It is clearly a combination giant, considering the associated and nonassociated gas production to date from the different reservoirs in the field. It may eventually prove to be a giant oil field for it needs only modest amounts of additional recovery to reach that status.

Table A.18

GIANT OIL FIELDS AND RECOVERABLE OIL RESOURCES OF DENMARK AS OF DECEMBER 31, 1975

(In millions of barrels)

Rank/Field	Discovery Year	1975 Production	Cumulative Production	Reserves	Total Recovery (With Range of Estimates)	
Others		1.2	3	193	200	
Denmark Total		1.2	3	193	200	(60-300)

NOTES: Past estimates of Danish reserves vary widely because of differences of opinion about the ultimate producibility of the tight chalk reservoirs in the *Dan* field offshore. The most recent opinions tend toward the pessimistic side. *Gorm*, roughly twice the size of *Dan*, should begin production in 1980.

Table A.19

GIANT OIL FIELDS AND RECOVERABLE OIL RESOURCES OF FRANCE AS OF DECEMBER 31, 1975

(In millions of barrels)

Rank/Field	Discovery Year	1975 Production	Cumulative Production	Reserves	Total Recovery (With Range of Estimates)	
Others		7.5	333	92	425	
France Total		7.5	333	92	425	(400-425)

NOTE: Nearly half of cumulative French production has come from the *Parentis* field. As the total figures indicate, it does not approach giant status.

Table A.20

GIANT OIL FIELDS AND RECOVERABLE OIL RESOURCES OF THE FEDERAL REPUBLIC OF GERMANY AS OF DECEMBER 31, 1975

(In millions of barrels)

Rank/Field	Discovery Year	1975 Production	Cumulative Production	Reserves	Total Recovery (With Range of Estimates)	
Possible Giants						
Emlichheim[a]	1944	1.2	30	10	40 (low?)	
Others		40.7	1,086	374	1,460	
FR Germany Total		41.9	1,116	384	1,500	(1,500-1,650)

[a]Combined with *Schoonebeek* (The Netherlands) for a total of 275 million barrels.

NOTES: *Schoonebeek-Emlichheim,* with approximately 1500 million barrels of 25° API oil-in-place originally, may become a marginal giant with a planned steam injection operation. Most of the remainder of production in the Federal Republic is concentrated in fields of 25-100 million barrels each.

Table A.21

GIANT OIL FIELDS AND RECOVERABLE OIL RESOURCES OF GREECE AS OF DECEMBER 31, 1975

(In millions of barrels)

Rank/Field	Discovery Year	1975 Production	Cumulative Production	Reserves	Total Recovery (With Range of Estimates)	
Others		--	--	150	150	
Greece Total		--	--	150	150 (?)	(40-600)

NOTE: Following the discovery of *Prinos* offshore in 1974, there were assertions that it was a small giant. However, further tests proved to be disappointing and anticipations were scaled down accordingly.

Table A.22

GIANT OIL FIELDS AND RECOVERABLE OIL RESOURCES OF ITALY AS OF DECEMBER 31, 1975

(In millions of barrels)

Rank/Field	Discovery Year	1975 Production	Cumulative Production	Reserves	Total Recovery (With Range of Estimates)	
Combination Giants						
267. *Malossa*	1973	0.5	1	324	325	(300-325)
Possible Giants						
*Rospo/Nosello***	1975	--	--			
Others		6.5	202	98	300	
Italy Total		7.0	203	422	625	(550-850)

NOTES: *Malossa* is a deep gas-condensate (51° API) field with estimated gas reserves of 1.75 Tcf. Because a majority of its energy-equivalent reserves are liquids, it is included here. *Rospo-Nosello* is a recent offshore discovery with an estimated 2900 million barrels of 15°-16° API, high sulfur oil-in-place. The combination of heavy oil and offshore location means that it will be very costly to produce. Most of the remainder of Italian production comes from the *Gela* and *Ragusa* fields in Sicily, two large but clearly non-giant heavy oil fields. Successful thermal recovery operations in both fields could increase their reserves substantially.

Table A.23

GIANT OIL FIELDS AND RECOVERABLE OIL RESOURCES OF THE NETHERLANDS AS OF DECEMBER 31, 1975

(In millions of barrels)

Rank/Field	Discovery Year	1975 Production	Cumulative Production	Reserves	Total Recovery (With Range of Estimates)		
Possible Giants							
Schoonebeek[a]	1943	5.6	170	65	235	(low?)	
Others		4.1	131	59	190		
The Netherlands Total		9.7	301	124	425	(low?)	(400-550)

[a]Combined with *Emlichheim* (Federal Republic of Germany) for a total of 275 million barrels.

NOTES: *Schoonebeek-Emlichheim*, with approximately 1500 million barrels of 25° API oil-in-place originally, may become a marginal giant with a planned steam injection operation. Most of the remaining production in The Netherlands is in two other heavy oil fields, *Ijsselmonde/Ridderkerk* and *Wassenaar-Meyendel*.

Table A.24

GIANT OIL FIELDS AND RECOVERABLE OIL RESOURCES OF NORWAY AS OF DECEMBER 31, 1975

(In millions of barrels)

Rank/Field	Discovery Year	1975 Production	Cumulative Production	Reserves	Total Recovery (With Range of Estimates)		
Giants							
47. *Statfjord*[**a]	1974	--	--	2,260	2,260	(low?)	(2,260-3,430)
100. *Ekofisk*[**]	1969	69.1	108	1,117	1,225	(low?)	(875-7,000)
206. *Murchison*[**b]	1975	--	--	60	60	(?)	
Combination Giants							
259. *Eldfisk*[**]	1970	--	--	420	420	(low?)	(380-1,000)
262. *Valhall*[**]	1975	--	--	400	400	(low?)	(350-500)
270. *Albuskjell*[**]	1972	--	--	245	245	(?)	(60-250)
Possible Giants							
W. Ekofisk[**]	1970	--	--	200	200	(?)	(170-1,000)
Tor[**]	1970	--	--	200	200	(?)	(125-375)
Others		--	--	490	490		
Norway Total		69.1	108	5,392	5,500		(5,000-7,000)
Post-1975 Discoveries							
Block 33/9-7&8	1976	--	--				

[a]Also in the United Kingdom for a combined total of 2570 million barrels.

[b]Also in the United Kingdom for a combined total of 560 million barrels.

NOTES: Because all of the Norwegian fields are recent discoveries that have little or no production records, estimates of both reserves and oil-in-place in each of the fields have varied widely from year to year, even from the same source. The estimates used here are based on those made by the Norwegian Petroleum Directorate, modified by information from the operators of the various fields. They should be considered tentative, pending additional information about reservoir performance that will be forthcoming during the next several years.

Statfjord is the largest field discovered to date in the North Sea. The estimate here is based on 50% recovery of the oil-in-place. Estimates of both oil-in-place and the recovery factor have varied. The estimate for *Ekofisk* is based on only 23% recovery. With successful gas injection or other secondary recovery possibilities, it could reach the 1500-1800 million barrels claimed for it by other recent sources.

Eldfisk and *Valhall* could easily prove to be oil giants as the estimates of recoverable oil for each are based on recovery factors of 13% and 25%, respectively. *Valhall* is planned to peak at 85 Mb/d and 180 MMcfd. *Albuskjell*, with an estimated 1.8 Tcf gas reserves, is a marginal combined giant, including gas liquids. *W. Ekofisk* and *Tor* could become marginal combined giants with additional recovery (their current oil and oil-equivalent reserves amount to 290 and 310 million barrels, respectively). As of the end of 1977, *W. Ekofisk* was producing at a rate of 80 Mb/d and 400 MMcfd indicating that it might be larger than indicated here. The two wells drilled in the structure north of *Statfjord* in *Block 33/9* have tested at promising rates.

Table A.25

GIANT OIL FIELDS AND RECOVERABLE OIL RESOURCES OF SPAIN AS OF DECEMBER 31, 1975

(In millions of barrels)

Rank/Field	Discovery Year	1975 Production	Cumulative Production	Reserves	Total Recovery (With Range of Estimates)	
Others		13.5	38	262	300	
Spain Total		13.5	38	262	300	(300-500)

NOTE: Although several encouraging discoveries have been made offshore Spain since 1970, no giants have been discovered.

Table A.26

GIANT OIL FIELDS AND RECOVERABLE OIL RESOURCES OF THE UNITED KINGDOM AS OF DECEMBER 31, 1975

(In millions of barrels)

Rank/Field	Discovery Year	1975 Production	Cumulative Production	Reserves	Total Recovery (With Range of Estimates)		
Giants							
47. *Statfjord***[a]	1974	--	--	310	310	(low?)	(310-470)
58. *Brent***	1971	--	--	2,000	2,000	(?)	(1,400-2,250)
70. *Forties***	1970	2.3	2	1,798	1,800	(low?)	(1,750-3,000)
111. *Ninian***	1974	--	--	1,100	1,100	(?)	(1,000-1,200)
178. *Piper***	1973	--	--	700	700	(low?)	(600-1,250)
190. *Brae***	1975	--	--	600	600	(?)	(200-2,000)
200. *Dunlin***	1973	--	--	585	585	(?)	(350-1,250)
206. *Murchison***[b]	1975	--	--	500	500	(?)	(300-700)
208. *Thistle***	1973	--	--	550	550	(?)	(350-1,000)
222. *Beryl***	1972	--	--	500	500	(?)	(380-800)
Probable Giants							
*Alwyn***	1973	--	--	425	425	(?)	(400-625)
*Andrew***	1974	--	--	450	450	(?)	(450-600)
*Claymore***	1974	--	--	405	405	(low?)	(360-500)
*North Cormorant***	1972	--	--	400	400	(?)	(400)
*Fulmar***	1975	--	--	400	400	(?)	
*Magnus***	1974	--	--	450	450	(?)	(400-1,000)
Possible Giants							
*Crawford***	1973	--	--	300	300	(?)	(200-500)
*Tartan***	1974	--	--	400	400	(?)	(300-500)
*Tern***	1975	--	--	300	300		(300+)
Others		6.6	24	1,901	1,925		
United Kingdom Total		8.9	26	14,074	14,100	(low?)	(10,000-16,000)
Post-1975 Discoveries							
*Renée***	1976	--	--	375	375	(?)	(375)
*Thelma/Toni***	1976	--	--	600	600	(?)	(600-750)

[a]Also in Norway for a combined total of 2570 million barrels.

[b]Also in Norway for a combined total of 560 million barrels.

NOTES: Estimates of total recovery from the major fields in the United Kingdom are uncertain because all of them are recent discoveries, most have not yet begun production, and some are not even fully delineated. Because most of the giant fields are marginal (500-600 million barrels), it is possible that a few may be downgraded in the future. On the other hand, additions to the list of confirmed U.K. giants are highly probable. The estimates already assume relatively high recovery factors (30%-45%), reflecting the fact that most of the fields will have pressure maintenance from the start. However, there is some potential for reserve growth as reservoir performance and per-well productivity in most of the fields that have begun production are turning out to be better than anticipated. Changes may not always be positive, as there have also been a few unpleasant surprises.

Table A.26--continued

Brent is anticipated to peak at 550 Mb/d (including 100 Mb/d NGL). The estimate for *Forties* is based on 40% recovery of an estimated 4400 million barrels of oil-in-place. The recent increase in planned peak production from 400 Mb/d could foreshadow an increase in recovery as well. *Ninian* could peak at 400 Mb/d, increasing estimated recovery moderately. The estimate for *Piper* is based on 45% recovery. With the recent increase in peak planned production from 250 to 320 Mb/d, 50% or more might be attainable. The estimates for *Brae* have fluctuated with the completion of each extension. Enough wells have now been drilled to confirm its giant status, but its ultimate potential is still rather uncertain. Estimates of *Dunlin's* potential show a wide variation; the estimate used here is the most recent. *Murchison* is rumored to have an estimated peak production of 150 to 200 Mb/d. *Thistle* is planned to peak at 200 Mb/d. *Beryl* appears to be a minimal oil giant or combination giant.

The complex *Alwyn* structure has yet to be fully assessed. *Andrew* is still too new a discovery for its status to be definitely determined. *Claymore*, with at least 1500 million barrels of oil-in-place, could become a giant with successful secondary recovery. *North Cormorant* is expected to peak at 180 Mb/d. The *Fulmar* discovery was just announced in 1977 and is said to have an expected output of 150 Mb/d, a level of production generally indicating a minimal North Sea giant. *Magnus* appears to be a significant discovery, but its complex geology and 600 foot water depths may inhibit recovery.

Crawford could prove to be a combination giant. *Tartan* and *Tern* are also marginal giant candidates, with the more recent estimates suggesting a 300-400 million barrel potential in each. *Renée* and *Thelma/Toni* are both sizeable fields, with *Thelma/Toni* being a clear giant. As an illustration of the fact that reserve changes can occur in either direction, there are seven fields in the United Kingdom--*Argyll, Buchan, Cormorant, Heather, Hutton, Maureen,* and *Montrose*--that have been asserted to be giants by at least one source shortly after their discovery, but that are now known not to merit that status.

SOURCES:

1. Department of Energy, United Kingdom, *Development of the Oil and Gas Resources of the United Kingdom, 1977*, London, April 1977.

2. A. N. Thomas, P. J. Walmsley, and D. A. Jenkins, "Forties Field, North Sea," *American Association of Petroleum Geologists Bulletin*, Vol. 58, No. 3, March 1974, pp. 396-406.

3. T. M. Thomas, "Search for Hydrocarbons in Shelf Seas of Northwest Europe: Progress and Prospects," *American Association of Petroleum Geologists Bulletin*, Vol. 59, No. 4, April 1975, pp. 573-617.

4. J. M. Watson and C. A. Swanson, "North Sea--Major Petroleum Province," *American Association of Petroleum Geologists Bulletin*, Vol. 59, No. 7, July 1975, pp. 1098-1112.

5. J. J. Williams, D. C. Conner, and K. E. Peterson, "Piper Oil Field, North Sea Fault Block Structure with Upper Jurassic Beach/Bar Reservoir Sands," *American Association of Petroleum Geologists Bulletin*, Vol. 59, No. 9, September 1975, pp. 1585-1601.

Table A.27

GIANT OIL FIELDS AND RECOVERABLE OIL RESOURCES OF YUGOSLAVIA AS OF DECEMBER 31, 1975

(In millions of barrels)

Rank/Field	Discovery Year	1975 Production	Cumulative Production	Reserves	Total Recovery (With Range of Estimates)	
Others		27.4	325	375	700	
Yugoslavia Total		27.4	325	375	700	(700)

NOTES: No data on individual fields in Yugoslavia could be located for this report. Descriptions of Yugoslavian production, however, indicate that it comes from a number of small- to medium-size fields, none of which approach giant status.

Table A.28

KNOWN RECOVERABLE CRUDE OIL RESOURCES OF EASTERN EUROPE AND THE SOVIET UNION BY FIELD CLASSIFICATION AS OF DECEMBER 31, 1975

(In millions of barrels)

Field Classification	Number of Fields	1975 Production	Cumulative Production	Reserves	Total Recovery
Giant fields	34	2,361.5	28,061	35,379	63,440
Combination giants	--	--	--	--	--
Probable giants	--	--	--	--	--
Possible giants[a]	14	49.0(+)	1,084(+)	1,166(+)	2,250(+)
Other fields[a]		1,244.1(-)	22,144(-)	14,566(-)	36,710(-)
Total		3,654.6	51,289	51,111(+)	102,400(+)

[a]The amounts in possible giants are understated because of a lack of information on several fields in both Rumania and the Soviet Union. Because the amounts for some of these fields are included in other fields, those amounts are slightly overstated.

NOTES: Giant oil fields are not as predominant in Eastern Europe and the Soviet Union as they are in the world as a whole. Known and potential giant fields in the region contain only about two-thirds of the known recoverable oil, the known giants having 63.44 billion barrels (62.0%) and the potential giants having at least 2.25 billion barrels (2.2%). Other fields have somewhat less than 36.71 billion barrels (35.8%). No information was available from which reliable estimates of the amount in large fields could be calculated. Sketchy information from Soviet sources suggests that this amount is likely to be between 33% and 50% of the amount in other fields.

As is true elsewhere in the world, giant fields in Eastern Europe and the Soviet Union promise to provide a larger share of future production. Known giants contain an estimated 35.38 billion barrels of reserves (69.2% of the total), and potential giants contain at least 1.17 billion barrels (2.3%). Non-giant fields have only 14.57 billion barrels of reserves (28.5%).

The available information on giant and non-giant oil fields (particularly the latter) in Eastern Europe and the Soviet Union is inadequate for high-confidence estimates of reserves and total recovery. However, the information available in translation is clearly increasing, particularly about giant fields.

SOURCES:

1. *European Petroleum Year Book, 1976,* Otto Vieth Verlag, Hamburg, 1977.
2. R. E. King, "Petroleum Exploration and Production in Europe in 1975," *American Association of Petroleum Geologists Bulletin,* Vol. 60, No. 10, October 1976, pp. 1704-1766 (also previous years).

Table A.29

GIANT OIL FIELDS AND RECOVERABLE OIL RESOURCES OF ALBANIA AS OF DECEMBER 31, 1975

(In millions of barrels)

Rank/Field	Discovery Year	1975 Production	Cumulative Production	Reserves	Total Recovery (With Range of Estimates)		
Others		15.0[E]	180[E]	140	320	(?)	
Albania Total		15.0[E]	180[E]	140	320	(?)	(320)

NOTES: Published information about Albanian oil fields and production generally consists of approximate estimates based on scanty information. It is clear, however, that there are no known giant fields in the country.

Table A.30

GIANT OIL FIELDS AND RECOVERABLE OIL RESOURCES OF BULGARIA AS OF DECEMBER 31, 1975

(In millions of barrels)

Rank/Field	Discovery Year	1975 Production	Cumulative Production	Reserves	Total Recovery (With Range of Estimates)		
Others		0.9[E]	40[E]	15	55		
Bulgaria Total		0.9[E]	40[E]	15	55	(?)	(55)

NOTE: Bulgarian production comes from a handful of small fields.

Table A.31

GIANT OIL FIELDS AND RECOVERABLE OIL RESOURCES OF CZECHOSLOVAKIA AS OF DECEMBER 31, 1975

(In millions of barrels)

Rank/Field	Discovery Year	1975 Production	Cumulative Production	Reserves	Total Recovery (With Range of Estimates)		
Others		1.0	36[E]	14	50		
Czechoslovakia Total		1.0	36[E]	14	50	(?)	(50)

NOTE: Only several small fields have been discovered to date in Czechoslovakia.

Table A.32

GIANT OIL FIELDS AND RECOVERABLE OIL RESOURCES OF EAST GERMANY AS OF DECEMBER 31, 1975
(In millions of barrels)

Rank/Field	Discovery Year	1975 Production	Cumulative Production	Reserves	Total Recovery (With Range of Estimates)
Others		1.6[E]	11[E]	24	35
East Germany Total		1.6[E]	11[E]	24	35

NOTE: East German production comes from two small fields.

Table A.33

GIANT OIL FIELDS AND RECOVERABLE OIL RESOURCES OF HUNGARY AS OF DECEMBER 31, 1975
(In millions of barrels)

Rank/Field	Discovery Year	1975 Production	Cumulative Production	Reserves	Total Recovery (With Range of Estimates)
Others		15.0[E]	326[E]	224	550
Hungary Total		15.0[E]	326[E]	224	550 (550-1,000)

NOTE: There are two sizeable oil fields in Hungary, neither of which approaches giant status.

Table A.34

GIANT OIL FIELDS AND RECOVERABLE OIL RESOURCES OF POLAND AS OF DECEMBER 31, 1975
(In millions of barrels)

Rank/Field	Discovery Year	1975 Production	Cumulative Production	Reserves	Total Recovery (With Range of Estimates)
Others		4.1[E]	336[E]	54[E]	390
Poland Total		4.1[E]	336[E]	54[E]	390 (?) (380-425)

NOTE: Oil production has occurred in Poland since the 1860s, but only several small- and medium-size oil fields have been discovered in the country.

Table A.35

GIANT OIL FIELDS AND RECOVERABLE OIL RESOURCES OF RUMANIA AS OF DECEMBER 31, 1975

(In millions of barrels)

Rank/Field	Discovery Year	1975 Production	Cumulative Production	Reserves	Total Recovery (With Range of Estimates)
Giants					
156. *Moreni-Gura Ocnitei*	1900	3.5[E]	758[E]	42[E]	800 (low?)
Possible Giants					
Boldesti					
Bustenari			*(Information not available)*		
Floresti-Tintea					
Others (includes possible)		105.2	2,702[E]	1,498[E]	4,200 (?)
Rumania Total		108.7[E]	3,460[E]	1,540	5,000 (?) (4,700-5,200)

NOTES: Information about production in the individual fields of Rumania is not available in recent Western sources. Given the absence of more recent information, the estimates of total recovery for *Moreni-Gura Ocnitei* are taken from Halbouty, et al. The production and reserve estimates are extrapolated assuming that production in the field is declining. Because of the discovery and development of deeper zones of production and of the application of additional recovery techniques, the estimate of total recovery could be larger. These two factors may also have proved sufficient to push other large oil fields in the Ploesti District, such as *Boldesti, Bustenari,* and *Floresti-Tintea* into the giant category.

Table A.36

GIANT OIL FIELDS AND RECOVERABLE OIL RESOURCES OF THE SOVIET UNION AS OF DECEMBER 31, 1975
(In millions of barrels)

Rank/Field	Discovery Year	1975 Production	Cumulative Production	Reserves	Total Recovery (With Range of Estimates)		
Giants							
11. *Samotlor*	1966	635.0	1,627	11,373	13,000	(?)	(11,000-15,000)
13. *Romashkino*	1948	580.0[E]	8,800[E]	3,600	12,400	(?)	(13,000-15,000)
45. *Arlan*	1955	100.0[E]	1,125[E]	1,675	2,800	(?)	(4,100-4,600)
50. *Uzen*	1961	115.0[E]	670[E]	1,830	2,500	(?)	(3,500-5,300)
51. *Balakhany-Sabunchi-Ramani*	1896	5.0[E]	2,350[E]	50	2,400	(?)	(2,400)
53. *Malgobek-Vosnesensko-Aliyurt*	1915	55.0[E]	2,000[E]	300	2,300	(?)	(to 3,000)
54. *Ust-Balyk*	1961	110.0[E]	690[E]	1,610	2,300	(?)	(2,850-5,100)
65. *Tuymazy*	1937	40.0[E]	1,660[E]	240	1,900	(?)	(2,200-3,600)
73. *Mamontovo*	1965	70.0[E]	200[E]	1,550	1,750	(?)	(1,750-4,000)
78. *Mukhanovo*	1945	55.0[E]	1,195[E]	355	1,550	(?)	(1,500-1,550)
88. *Kotur-Tepe*	1956	70.0[E]	720[E]	740	1,460	(?)	(1,460-4,000)
92. *Novo-Elkhov*	1955	60.0[E]	630[E]	670	1,300	(?)	(3,000-3,200)
97. *Shkapovo*	1953	40.0[E]	950[E]	300	1,250	(?)	(1,250-1,650)
99. *Sovetskoye*	1962	50.0[E]	265	985	1,250	(?)	(1,600-4,200)
101. *Bibi Eybat**	1848	10.0[E]	1,110[E]	90	1,200	(?)	(to 2,000)
115. *Neftyany Kamni**	1949	28.0[E]	850[E]	200	1,050	(?)	(1,200-2,500)
133. *Usin*	1963	30.0[E]	52[E]	948	1,000	(?)	(600-2,000)
134. *Vozey*	1972	--	--	1,000	1,000	(?)	
139. *Pravdinsk*	1964	49.0[E]	201[E]	749	950	(?)	(1,200-1,500)
147. *Baku Archipelago**	1963	34.0[E]	195[E]	665	860	(?)	(860-900)
152. *Bolshoye-Chernogor*	1970	--	--	800	800	(?)	
157. *Pokachev*	1970	--	--	800	800	(?)	
159. *Yarino*	1956	40.0[E]	385[E]	415	800	(?)	(510-640)
162. *Kuleshova*	1958	40.0[E]	485[E]	275	760	(low?)	(760)
171. *West Surgut*	1962	40.0[E]	250	480	730	(?)	(730-2,000)
184. *Megion*	1961	30.0[E]	190[E]	440	630	(?)	(730-890)
187. *Agan*	1966	12.0[E]	22[E]	578	600	(?)	
192. *Fedorov*	1971	30.0[E]	46[E]	554	600	(?)	(1,000-5,000)
193. *Karachukhur-Zykh*	1928	10.0[E]	510[E]	90	600	(?)	(600)
198. *North Varyegan*	1971	--	--	600	600	(?)	
228. *Kholmogor*	1973	--	--	500	500	(?)	(1,000-5,000)
242. *Vat'yegan*	1971	--	--	500	500	(?)	
243. *Zhetybai*	1960	20.0[E]	125[E]	375	500	(?)	(to 1,100)

Table A.36--continued

Rank/Field	Discovery Year	1975 Production	Cumulative Production	Reserves	Total Recovery (With Range of Estimates)		
Possible Giants							
Cheleken[*]	1913						(640)
Glynsko-Rozybyshev	1958						(310)
Karazhanabas	1974	--	--				
Novoportov	1964	--	--				
Oktyabr'skoye	1913	4.0^E	420^E	30	450	(?)	(520)
North Pokur	1964	--	--	300	300		
Priluki	1959	15.0^E	170^E	130	300	(?)	(660)
Russkoye	1968	--	--				
Salym	1963	1.0	1				(to 10,000)
Surakhany	1907						
Starogrosni	1893	4.0^E	380^E	20	400	(?)	(650)
Varyegan	1970	--	--	400	400	(?)	
Vatin	1961	25.0^E	113^E	287	400	(?)	
Others		$1,101.3^E$	$18,514^E$	12,596	31,110	(?)	
Soviet Union Total		3,508.3	$46,900^E$	49,100+	96,000+	(?)	(87,000-127,000)

NOTES: Because the Soviet Union does not publish reserve statistics, precise estimates of reserves are not available. Reasonable estimates of Soviet reserves by field and countrywide totals can be pieced together from information about cumulative production, production plans and trends (of both oil and water), and field characteristics. The estimates shown here for the individual giant fields are in many cases lower than estimates published in other sources. These other estimates were based primarily on early overly optimistic evaluations of field potential by the Soviets. However, the waterfloods in the fields developed to date are proving to be less efficient than anticipated, and delineation drilling in some of the newer West Siberian fields is resulting in a scaling-down of early reserve estimates based on a handful of wells.

Production in *Samotlor* (Western Siberia) is expected to peak at 2.6 MMb/d from 1977 to 1980 and decline thereafter. (Ultimate production in each of the Western Siberian fields is likely to be twelve to fifteen times peak annual production.) Production in *Romashkino* (Tataria) is now beginning to decline sharply after producing at a level of 1.6 MMb/d from 1970 to 1976. The design of the waterflood in the lenticular sands of *Arlan* (Bashkiria) resulted in a loss of reserves exceeding 30%. The injection of 30°C untreated water into *Uzen* (Kazakhstan), with its highly paraffinic oil, appears to have resulted in substantial reservoir damage. *Balakhany-Sabunchi-Ramani* (Azerbaijan), one of the oldest fields in the Soviet Union, is nearly depleted. Production in *Malgobek-Vosnesensko-Aliyurt* (Chechen-Ingush) is now dropping very rapidly. Although production in the main zone of *Ust-Balyk* is now declining, it has some recent extensions that have yet to be fully developed. After an extensive waterflood, production in *Tuymazy* (Bashkiria) has begun to decline. Production in *Mamontovo* (Western Siberia) has just reached its planned peak of 240 Mb/d. Because its oil is heavier than that in most of the Western Siberian fields (27° API versus 33°-35°), it is likely to have a slightly longer producing life. *Mukhanovo* (Kuibyshev) is one of the few Volga-Ural giants that is likely to meet expectations. Production in *Kotur-Tepe* (Turkmenia) is now beginning to decline after a plateau of several years at 200 Mb/d. *Novo-Elkhov* (Tataria) will be declining by 1980.

The waterflooding pattern used in *Shkapovo* (Bashkiria) resulted in a reserve loss of about 15%. Given its anticipated 240 Mb/d peak, the potential of *Sovetskoye* (Western Siberia) appears to be substantially less than indicated by other sources. *Bibi Eybat* (Azerbaijan), the first giant field discovered in the world, is nearly depleted. *Pravdinsk* (Western Siberia) is expected to peak at 160 Mb/d within a few years. Production in *Neftyany Kamni* (Azerbaijan) has declined sharply from a 150 Mb/d peak in 1970. The estimates for *Usin* and *Vozey* (both Komi A.S.S.R.) could be conservative given the area (around 100,000 acres each) and amplitude of each structure. Because most of the producing zones in the *Baku Archipelago* (aka *Sangachaly-Duvannyi-Bulla More* in Azerbaijan) are deep, they may not be fully developed given the inefficiency of Soviet deep drilling. However, the possibility of reservoir damage in the field has been mentioned. Production in *Pravdinsk*,

Table A.36--continued

Bolshoye-Chernogor, and *Pokachev* (all in Western Siberia) is expected to peak at 160 Mb/d. *Bolshoye-Chernogor* is considered by some sources to be an extension of *Samotlor*. *Yarino* (Perm) is likely to be larger than indicated in other sources, given its increasing production during the early 1970s.

Kuleshova (Kuibyshev), the last giant field to be discovered in the Volga-Ural region, may prove to be somewhat larger than indicated here. *West Surgut* and *Megion*, two of the earliest giant discoveries in Western Siberia, are reported to be declining (although there are conflicting recent reports about the potential of the latter). *Fedorov* (Western Siberia) was once thought to have a potential of 160 Mb/d, but expectations have been scaled down. *Karachukhur-Zykh* (Azerbaijan) is nearly depleted. *Agan* and *North Varyegan* (Western Siberia) are expected to reach 120 Mb/d by 1980. *Kholmogor* and *Vat'yegan* (Western Siberia) are expected to peak at 100 Mb/d (though the former was once expected to have considerably greater potential). *Zhetybai* (Kazakhstan) shares both the same characteristics and the same problems as *Uzen*. It may be only a combination giant.

Cheleken (Turkmenia), *Oktyabr'skoye* (Chechen-Ingush), *Priluki* (Ukraine), and *Starogrosni* (Chechen-Ingush) were indicated as giants by Halbouty et al. This status is not substantiated by production trends since 1970 and the latest Soviet plans for the areas in which they are located. The status of *Glynsko-Rozybyshev* (Ukraine), indicated as a combination giant by Halbouty et al., could not be determined in any reliable manner. *Karazhanabas* (Kazakhstan) is a recent discovery of heavy, high-sulfur oil in a large structure that could prove to be a giant. *Novoportov* and *Russkoye* are two large fields in northern Western Siberia with reservoirs of either heavy (18° API) or highly paraffinic oil. As both are overlain by 1000 to 2000 feet of permafrost, development, if it proves to be feasible, will be very expensive. *North Pokur* (Western Siberia) could be a combination giant. *Salym* (Western Siberia) has been touted as a major discovery for over a decade, but as yet no production targets for it have been published, probably because test wells in the Lower Cretaceous were not on average very productive. A deeper Jurassic pool discovery in 1974 could confirm its giant status. *Surakhany* (Azerbaijan) could be a nearly depleted combination giant. Some very sketchy information indicates that its cumulative production could be about 300-350 million barrels. *Varyegan* (Western Siberia), with a planned peak production of 80 Mb/d and a sizeable gas cap, could be a combination giant. *Vatin* could be a minimal giant if any of its new-pool discoveries during the early 1970s are of significant size. *South Cheremshanka* (Western Siberia), announced as a giant shortly after its discovery in 1969, has had almost nothing said about it since, indicating that it belongs in the category of erstwhile giants.

The estimate for total U.S.S.R. reserves includes A, B, and some C_1 reserves (a combination that is roughly analogous to Western definitions of proved and probable reserves). Given Soviet statements about declining reserve-production ratios and recent declines in reserves, significantly higher estimates of proved and probable reserves are quite unlikely. This does not imply, however, that there is little potential for further discoveries.

SOURCES:

1. R. W. Campbell, *The Economics of Soviet Oil and Gas*, The Johns Hopkins Press, Baltimore, Md., 1968.

2. R. W. Campbell, *Trends in the Soviet Oil and Gas Industry*, The Johns Hopkins Press, Baltimore, Md., 1976.

3. Central Intelligence Agency, *Prospects for Soviet Oil Production*, ER 77-10270, Washington, D.C., April 1977.

4. Central Intelligence Agency, *Prospects for Soviet Oil Production: A Supplemental Analysis*, ER 77-10425, Washington, D.C., July 1977.

5. J. W. Clarke et al., *Petroleum Geology of the West Siberian Basin and a Detailed Description of the Samotlor Oil Field*, U.S. Geological Survey Open-File Report 77-871, Reston, Va., 1977.

6. *Petroleum Geology* (a monthly digest of articles from the Soviet oil and gas literature in English translation).

Table A.37

KNOWN RECOVERABLE CRUDE OIL RESOURCES OF AFRICA BY FIELD CLASSIFICATION AS OF DECEMBER 31, 1975
(In millions of barrels)

Field Classification	Number of Fields	1975 Production	Cumulative Production	Reserves	Total Recovery
Giant fields	30	1,062.6	13,831	36,259	50,090
Combination giants	1	3.9	249	211	460
Probable giants	6	106.1	1,132	1,618	2,750
Possible giants	9	93.6	687	1,933	2,620(+)
Other fields		541.4	4,646	15,079	19,725
Total		1,807.6	20,545	55,100(+)	75,645(+)

NOTES: Nearly all of the known recoverable crude oil resources of Africa are in giant or large fields. The 31 known giant fields have 50.55 billion barrels (66.8% of the total). The 15 potential giants contain at least 5.37 billion barrels (7.1%). Of the 19.73 billion barrels (26.1%) in non-giant fields, approximately 14.18 billion are in 89 large fields, most of which are in Nigeria.

The distribution of reserves by size of field in Africa is quite similar to the distribution of total recovery. Known giants have 36.26 billion barrels (65.8%), potential giants have at least 3.55 billion barrels (6.4%), and other fields have 15.08 billion barrels (27.4%).

The data available on African giants are of reasonable quality. The major uncertainties relate to reserve additions which may be forthcoming from the implementation of secondary and enhanced recovery operations and the gas reserves of several potential combination giants.

SOURCES:

1. *Arab Oil and Gas Directory, 1976-1977,* Arab Petroleum Research Center, Beirut, 1976.
2. P. Biro, "Petroleum Developments in Central and Southern Africa in 1975," *American Association of Petroleum Geologists Bulletin,* Vol. 60, No. 10, October 1976, pp. 1813-1864 (also previous years).
3. M. A. Nicod, "Petroleum Developments in North Africa in 1975," *American Association of Petroleum Geologists Bulletin,* Vol. 60, No. 10, October 1976, pp. 1767-1812 (also previous years).

Table A.38

GIANT OIL FIELDS AND RECOVERABLE OIL RESOURCES OF ALGERIA AS OF DECEMBER 31, 1975

(In millions of barrels)

Rank/Field	Discovery Year	1975 Production	Cumulative Production	Reserves	Total Recovery (With Range of Estimates)		
Giants							
25. *Hassi Messaoud*	1956	210.0	2,178	4,822	7,000	(?)	(3,500-11,000)
112. *Zarzaitine*	1958	34.8	540	560	1,100	(?)	(850-1,300)
166. *El Borma*[a]	1967(1964)	9.0	50	200	250		
226. *Gassi Touil*	1963	27.4	70	430	500	(?)	(315-500)
Combination Giants							
253. *Rhourde El Baguel*	1962	3.9	249	211	460	(?)	(315-500)
Possible Giants							
Edjeleh	1955	6.9	165	135	300	(?)	(250-300)
El Agreb	1959	12.0[E]	150[E]	200	350	(?)	
Haoud Berkaoui	1965	11.8	103	147	250	(low?)	
Others		2.5	466	1,324	1,790		
Algeria Total		318.3	3,971	8,029	12,000		(10,000-14,000)

[a]Also in Tunisia for a combined total of 750 million barrels.

NOTES: The estimates given here for 1975 and cumulative production of crude oil exclude natural gas liquids from Algeria's giant gas fields. Inclusion of these amounts would increase the estimates by 10%-15%.

Hassi Messaoud has an estimated 20-25 billion barrels of oil-in-place. With enhanced recovery, which will be applied field-wide, the recovery factor could reach 50%, increasing the estimate used here by 3-5 billion barrels. Both water and gas injection began in *Zarzaitine* in the 1960s. Its gas cap also approaches giant dimensions. When it began production, the Algeria side of *El Borma* was stated to be eventually capable of 50 Mb/d or more. Only 25 Mb/d have been realized and little has been said about the Algerian side of the field in the oil industry press during the past five years. Production in *Gassi Touil* has been increasing steadily since it began in 1972. The field also has a gas cap of near-giant dimensions.

Rhourde El Baguel has a large gas cap. Its oil production has been restricted since 1973. Given the oil-in-place in *El Agreb* (1150 million barrels), it could prove to be a combination giant with enhanced recovery and potential gas production. *Edjeleh* could also be a combination giant with enhanced recovery. Shortly after *Haoud Berkaoui* began production, an ultimate goal of 150 Mb/d was indicated. Only 35 Mb/d were realized by 1976.

The super-giant (*Hassi R'Mel*) and giant (*Alrar* and *Rhourde Nous*) gas fields of Algeria contain several billion barrels of natural gas liquids. The planned increase in gas production from these fields by 1985 should result in a substantial increase in total Algerian production of petroleum liquids.

SOURCES:

1. O. Ali, "El Agreb-El Gassi Oilfields, Central African Sahara," *American Association of Petroleum Geologists Bulletin*, Vol. 59, No. 9, September 1975, pp. 1676-1684.
2. A. Balducchi and G. Pommier, "Cambrian Oil Field of Hassi Messaoud, Algeria," in M. T. Halbouty, ed., *Geology of Giant Petroleum Fields*, American Association of Petroleum Geologists Memoir 14, Tulsa, Okla., 1970, pp. 477-488.
3. W. F. Bishop, "Geology of Tunisia and Adjacent Parts of Algeria and Libya," *American Association of Petroleum Geologists Bulletin*, Vol. 59, No. 3, March 1975, pp. 413-450.

Table A.39

GIANT OIL FIELDS AND RECOVERABLE OIL RESOURCES OF ANGOLA AS OF DECEMBER 31, 1975

(In millions of barrels)

Rank/Field	Discovery Year	1975 Production	Cumulative Production	Reserves	Total Recovery (With Range of Estimates)		
Giants							
144. *Malongo West***	1969	34.5	134	766	900	(?)	
195. *Malongo North and South***	1966	16.8	149	451	600	(?)	
Others		6.3	78	222	300		
Angola Total		57.6	361	1,439	1,800		(1,650-1,720)

NOTES: The two *Malongo* fields are often combined in compilations of giant fields, even though they are two distinct structures. There is some possibility for reserve growth as the oil in some reservoirs of both fields is heavy gravity (20°-25° API).

Table A.40

GIANT OIL FIELDS AND RECOVERABLE OIL RESOURCES OF CONGO AS OF DECEMBER 31, 1975

(In millions of barrels)

Rank/Field	Discovery Year	1975 Production	Cumulative Production	Reserves	Total Recovery (With Range of Estimates)		
Possible Giants							
*Emeraude***	1969	12.3	46	374	420	(?)	(175-500)
*Loango***	1972	--	--	300	300	(?)	(150-700)
Others		0.1	6	1	7		
Congo Total		12.4	52	675	727		(525-2,500)

NOTES: *Emeraude* has an estimated 5-7 billion barrels of 24° API oil-in-place, but because of various technical and economic factors, recovery is expected to be less than 10%. With higher economic returns to the operators, it could eventually become a giant. The planned production capacity of *Loango* is 50 Mb/d. It could be a marginal giant. The basis for the high estimate of total recovery in the country as a whole is unclear, particularly because it is double the sum of the high estimates for the only two significant fields in the Congo.

Table A.41

GIANT OIL FIELDS AND RECOVERABLE OIL RESOURCES OF EGYPT AS OF DECEMBER 31, 1975

(In millions of barrels)

Rank/Field	Discovery Year	1975 Production	Cumulative Production	Reserves	Total Recovery (With Range of Estimates)		
Giants							
81. *El Morgan***	1965	31.6	474	1,026	1,500	(?)	(800-2,500)
130. *Ramadan***	1974	13.5	14	986	1,000	(?)	(500-1,000)
176. *July***	1973	16.1	23	677	700	(?)	(500-800)
Probable Giants							
Belayem	1955	11.5^E	275^E	200	475	(?)	
*Belayem Marine***	1961	13.0^E	223^E	202	425	(?)	
Others		23.3	509	341	850		
Egypt Total		109.0	1,518	3,432	4,950	(?)	(3,000-5,400)
Post-1975 Discoveries							
*Block 382***	1976	--	--				

NOTES: Estimates of Egyptian production and reserves are subject to several substantial uncertainties. Total recovery from the fields in the Gulf of Suez depends on the degree to which major water injection operations will be successful. Estimates of cumulative production of the Sinai fields occupied by the Israelis from 1967 to 1975 are plagued by a lack of reliable information and their future prospects are not yet apparent.

Water injection in *El Morgan* is being increased to boost production to 150-175 Mb/d. *Ramadan*, with a very thick producing section and a good natural water drive, has an estimated potential of 150-200 Mb/d. Water injection equipment was being installed in *July* in 1977, and its production is expected to reach 150 Mb/d.

Belayem and *Belayem Marine* were occupied and produced by the Israelis from 1967 to 1975. The two are separate fields, although they were combined by Halbouty et al. *Belayem* has heavy oil (22° API) and thus with any kind of enhanced recovery is likely to become a giant. The oil in *Belayem Marine* is lighter, but it may not yet be fully developed.

The discovery well of *Block 382* encountered two productive sections, each around 400 feet thick. Plans for it to reach 60 Mb/d by early 1978 have been announced. It is most likely a field of 200-300 million barrels' total recovery. *Amal*, a 1968 discovery, was counted as a giant (2500 million barrels) by Halbouty et al. This relatively thin (130 feet) reservoir of heavy (25°-26° API) oil was never developed and appears to have been abandoned. It is another of several erstwhile giants.

Table A.42

GIANT OIL FIELDS AND RECOVERABLE OIL RESOURCES OF GABON AS OF DECEMBER 31, 1975

(In millions of barrels)

Rank/Field	Discovery Year	1975 Production	Cumulative Production	Reserves	Total Recovery (With Range of Estimates)		
Probable Giants							
*Grondin***	1969	22.4	38	362	400	(low?)	(400-500)
Others		57.8	441	1,059	1,500		
Gabon Total		80.2	479	1,421	1,900	(low?)	(1,050-2,700)

NOTES: *Grondin* is expected to reach a peak production of 70 Mb/d. It could prove to be a marginal giant, particularly with improved economics as its exploration and development costs have been relatively high ($2-$3 per barrel). Gabon is another country for which the highest estimate of total recovery appears to be totally without foundation.

Table A.43

GIANT OIL FIELDS AND RECOVERABLE OIL RESOURCES OF LIBYA AS OF DECEMBER 31, 1975

(In millions of barrels)

Rank/Field	Discovery Year	1975 Production	Cumulative Production	Reserves	Total Recovery (With Range of Estimates)		
Giants							
24. *Sarir*	1961	53.5[E]	760[E]	6,440	7,200	(?)	(2,600-8,000)
36. *Amal*	1959	26.6	510	3,740	4,250	(?)	(4,250)
37. *Nasser*	1959	46.9	1,740	2,460	4,200	(?)	(2,200-4,200)
38. *Gialo*	1961	70.2	1,181	2,819	4,000	(?)	(2,000-4,200)
69. *Defa*	1960	48.3	394	1,406	1,800	(?)	(600-2,000)
71. *Intisar D*	1967	66.4	591	1,209	1,800	(?)	(1,150-1,300)
84. *Nafoora-Augila*	1965	44.8	744	756	1,500	(low?)	(1,300-1,500)
103. *Intisar A*	1967	21.3	524	676	1,200	(?)	(1,200-1,650)
104. *L-65*	1966	9.7[E]	122[E]	1,078	1,200	(?)	
107. *Waha*	1960	36.5	560	640	1,200	(low?)	(900-1,000)
119. *Raguba*	1961	18.6	429	611	1,040	(high?)	(1,000-1,040)
124. *Bu Attifel*	1968	40.0	142	858	1,000	(low?)	(600-1,000)
172. *Dahra-Hofra*	1959	7.0	445	255	700	(?)	(700-800)
189. *Bahi*	1968	16.1	190	410	600	(low?)	(350-400)
238. *Samah*	1961	11.3	248	252	500	(?)	(350-1,300)
Possible Giants							
Mabruk	1959	--	--				
Others		26.3	710	2,100	2,810		
Libya Total		543.5	9,290	25,710	35,000		(33,000-35,000)

NOTES: Production in Libya, as is true of most of the major oil-producing countries, is dominated by a few giant fields. The estimates of recoverable reserves in these giant fields and for the country as a whole are fairly reliable.

Sarir, the largest field in Libya, has an estimated 12-15 billion barrels of oil-in-place. Water injection has already begun; the estimate of total recovery here assumes roughly 55% recovery of 13 billion barrels in place. Water injection has already begun in *Amal*, which is incidentally quite distinct from *Nafoora-Augila* (Halbouty et al. combined the two). Production in both *Nasser* (formerly known as *Zelten*) and *Gialo* was cut back sharply in the early 1970s. Given recent production in *Defa* (190 Mb/d in early 1977), it seems likely to meet the earlier estimates suggested for it. *Intisar D* and *Intisar A* may have been overproduced in their first few years of production, but because pressure maintenance and enhanced recovery operations were applied early in their producing lives, a high recovery of the oil-in-place in each is likely. *Nafoora-Augila* peaked at over 400 Mb/d, an indication that it may have greater potential than indicated here. *L-65* is often lumped together with *Sarir*, even though they are distinct traps several miles apart. In the late 1960s, a capacity of 90 Mb/d was planned for it. Production at *Waha* has remained relatively steady between 110 and 150 Mb/d for nearly 15 years. *Raguba* production has declined 50% from its 1969 peak. The field may need the initiation of secondary recovery to reach the indicated total recovery. *Bu Attifel* was once expected to reach 275 Mb/d. By late 1976, production was up to 190 Mb/d, indicating that total recovery could well exceed 1 billion barrels. Production in *Bahi* in 1976 was nearly double the 1975 level. Published estimates of its reserves are clearly understated. Early estimates appear to have overestimated the potential of *Samah*. It is likely to be only a minimal giant.

Mabruk covers a sizeable area. However, it remains undeveloped because of its low permeability and consequently its low productivity per well and high development and production costs. Several offshore discoveries have been announced in 1975-1976, but pending resolution of the offshore boundary between Libya and Tunisia nothing but the most general information has been forthcoming about them. Several undeveloped fields

Table A.43--continued

in the Ghadames Basin of western Libya are asserted to have 1 billion barrels of reserves among them; none appear to be of giant stature.

SOURCES:

1. C. Rand, *Making Democracy Safe for Oil*, Little, Brown and Company, Boston, 1975.

2. J. M. Roberts, "Amal Field, Libya," in M. T. Halbouty, ed., *Geology of Giant Petroleum Fields*, American Association of Petroleum Geologists Memoir 14, Tulsa, Okla., 1970, pp. 438-448.

3. R. M. Sanford, "Sarir Oil Field, Libya-Desert Surprise," in M. T. Halbouty, ed., *Geology of Giant Petroleum Fields*, American Association of Petroleum Geologists Memoir 14, Tulsa, Okla., 1970, pp. 449-476.

4. C. E. Terry and J. J. Williams, "The Idris 'A' Bioherm and Oilfield, Sirte Basin, Libya--Its Commercial Development, Regional Paleocene Geologic Setting, and Stratigraphy," in P. Hepple, ed., *The Exploration for Petroleum in Europe and North Africa*, Institute of Petroleum, London, 1969.

5. J. J. Williams, "Augila Field, Libya: Depositional Environment and Diagenesis of Sedimentary Reservoir and Description of Igneous Reservoir," in R. E. King, ed., *Stratigraphic Oil and Gas Fields--Classification, Exploration Methods, and Case Histories*, Memoir 16, American Association of Petroleum Geologists, Tulsa, Okla., 1972, pp. 623-632.

Table A.44

GIANT OIL FIELDS AND RECOVERABLE OIL RESOURCES OF MOROCCO AS OF DECEMBER 31, 1975

(In millions of barrels)

Rank/Field	Discovery Year	1975 Production	Cumulative Production	Reserves	Total Recovery (With Range of Estimates)
Others		0.2	15	3	18
Morocco Total		0.2	15	3	18

NOTE: Despite extensive exploration, only a few very small oil fields have been found in Morocco to date.

Table A.45

GIANT OIL FIELDS AND RECOVERABLE OIL RESOURCES OF NIGERIA AS OF DECEMBER 31, 1975

(In millions of barrels)

Rank/Field	Discovery Year	1975 Production	Cumulative Production	Reserves	Total Recovery (With Range of Estimates)		
Giants							
154. *Jones Creek*	1967	29.0	211	589	800	(?)	(600-1,000)
174. *Imo River*	1959	22.6	280	420	700	(?)	(600-650)
179. *Yokri*	1968	31.7	190	510	700	(?)	(500-550)
182. *Meren***	1965	27.9	224	426	650	(?)	(500-600)
224. *Bomu*	1958	11.3	258	242	500	(?)	(550-625)
234. *Okan***	1964	14.3	232	268	500	(?)	(500-585)
Probable Giants							
Kokori	1961	18.6	192	258	450	(?)	(350-400)
Obagi	1966	19.3	106	294	400	(?)	
Olomoro-Afiesere-Eriemu-Oweh	1961	21.3	298	302	600	(?)	
Possible Giants							
*Delta South***	1965	16.1	129	221	350	(?)	(250-475)
Odidi	1967	17.9	68	282	350	(?)	
Others		421.8	2,404	9,596	12,000		
Nigeria Total		651.8	4,592	13,408	18,000		(15,000-24,500)

NOTES: Most of the fields discovered in the Niger Delta to date are large (100 to 500 million barrels each), but not giant fields. The contribution of giant fields to total recovery is the smallest proportion found in any major producing country. Unless there are a substantial number of undeveloped pools or there is a considerable potential for secondary recovery in the fields for which Shell/BP is the operator (which there may be, given that current recovery is less than 35% of oil-in-place), the upper range of estimates for total recovery in Nigeria appears to exaggerate Nigerian proved and probable reserves.

Jones Creek, with relatively heavy oil (25° API), produced more than 125 Mb/d in 1974. Total recovery in *Imo River*, *Yokri*, and *Meren* appears to be somewhat larger than indicated by other sources, given their production both before and after the 1975 slump in total Nigerian production. *Bomu* and *Okan*, on the other hand, appear to have slightly less potential than indicated by other sources.

Kokori could be a marginal combination giant field already. *Obagi*, if it has a sufficient volume of natural gas in nonassociated gas pools, could also be a combination giant. If *Olomoro* merges with *Afiesere*, *Eriemu*, and *Oweh*, as has been predicted, it will clearly be a giant field. It is indicated as a probable giant here for lack of definitive evidence that the merger will occur. By itself, *Olomoro* is a 350 million barrel field with a small chance of being a combination giant.

Delta South and *Odidi* are also potential combination giants, particularly the latter, which is indicated by B. D. Evamy et al. to have substantial gas resources. Because of a general lack of information on Nigerian gas resources by field, this compilation may not include all the potential combination giants and underrate those that are included. *Ekpe* and *Mbede*, each with around 300 million barrels' total recovery, could also have the potential to be marginal combination giants.

SOURCE: B. D. Evamy et al., "Hydrocarbon Habitat of Tertiary Niger Delta," *American Association of Petroleum Geologists Bulletin*, Vol. 62, No. 1, January 1978, pp. 1-39.

Table A.46

GIANT OIL FIELDS AND RECOVERABLE OIL RESOURCES OF TUNISIA AS OF DECEMBER 31, 1975

(In millions of barrels)

Rank/Field	Discovery Year	1975 Production	Cumulative Production	Reserves	Total Recovery (With Range of Estimates)		
Giants							
166. *El Borma*[a]	1964	14.9	224	276	500	(?)	(300-600)
Possible Giants							
*Ashtart***	1971	16.6	26	274	300	(?)	(250-800)
Others		3.0	17	133	150		
Tunisia Total		34.5	267	683	950	(?)	(1,100-3,000)

[a]Also in Algeria for a combined total of 750 million barrels.

NOTES: Total recovery in *El Borma* is likely to increase with the initiation of secondary recovery. It is also reported to contain substantial amounts of natural gas liquids. *Ashtart* has been asserted to be a giant in one source, but given its planned peak production (55 Mb/d), it seems unlikely that it could be one except in the most marginal way. The basis for the high range of estimates for total recovery in Tunisia as a whole is a mystery, as even the sum of the upper range of estimates for all the known fields is not even half of it.

Table A.47

GIANT OIL FIELDS AND RECOVERABLE OIL RESOURCES OF ZAIRE AS OF DECEMBER 31, 1975

(In millions of barrels)

Rank/Field	Discovery Year	1975 Production	Cumulative Production	Reserves	Total Recovery (With Range of Estimates)	
Others		0.1	*	300	300	0
Zaire Total		0.1	*	300	300	(150-500)

NOTE: Production in the two known fields of Zaire began in late 1975.

Table A.48

KNOWN RECOVERABLE CRUDE OIL RESOURCES OF THE MIDDLE EAST BY FIELD CLASSIFICATION AS OF DECEMBER 31, 1975

(In millions of barrels)

Field Classification	Number of Fields	1975 Production	Cumulative Production	Reserves	Total Recovery
Giant fields	77	6,920.4	83,562	414,843	498,405
Combination giants	2	--	--	740	740
Probable giants[a]	12	53.5	464	3,340(+)	3,804(+)
Possible giants[a]	38	32.9	309	2,017(+)	2,326(+)
Other fields		93.6	822	3,849	4,671
Total		7,100.4	85,157	424,789(+)	509,946(+)

[a]Because information is not available for many potential giant fields in the Middle East, the estimates shown here are clearly understated.

NOTES: Giant oil fields and known recoverable crude oil resources in the Middle East are practically synonymous. The 79 known giant fields contain at least 499.15 billion barrels, 97.9% of the known recoverable total. The 50 potential giants contain at least 6.13 billion barrels (1.2%), an amount that could easily be increased several times given more accurate information. Only 4.67 billion barrels (0.9%) are in other fields, of which 3.27 billion are in at least 16 large fields.

The distribution of reserves is similar to that of total recovery. Known giant fields have 415.58 billion barrels (97.8%) and potential giants have at least 5.36 billion (1.3%). Non-giant fields have only 3.85 billion (0.9%).

The data available on developed giant fields in the Middle East are reasonably accurate. The main uncertainty in these fields is the effect that secondary and enhanced recovery will have on reserve growth. There are many undeveloped potential giant fields in the Middle East for which almost no reasonably firm information about their economics or likely potential is available.

SOURCES:

1. *Arab Oil and Gas Directory, 1976-1977,* Arab Petroleum Research Center, Beirut, 1976.
2. Z. R. Beydoun and H. V. Dunnington, *The Petroleum Geology and Resources of the Middle East,* Scientific Press Ltd., Beaconsfield, England, 1975.
3. R. C. Hasson, J. F. Mason, and Q. M. Moore, "Petroleum Developments in Middle East Countries," *American Association of Petroleum Geologists Bulletin,* Vol. 60, No. 10, October 1976, pp. 1865-1899 (also previous years).
4. S. H. Longrigg, *Oil in the Middle East,* 3rd ed., Oxford University Press, London, 1968.
5. C. T. Rand, *Making Democracy Safe for Oil,* Little, Brown and Company, Boston, 1975.

Table A.49

GIANT OIL FIELDS AND RECOVERABLE OIL RESOURCES OF ABU DHABI AS OF DECEMBER 31, 1975

(In millions of barrels)

Rank/Field	Discovery Year	1975 Production	Cumulative Production	Reserves	Total Recovery (With Range of Estimates)		
Giants							
15. *Zakum***	1964	88.6	667	11,333	12,000	(?)	(1,000-2,000)
19. *Bu Hasa*	1962	170.2[E]	1,375[E]	7,625	9,000	(?)	(2,400-3,000)
28. *Asab*	1965	123.5[E]	213[E]	5,787	6,000	(?)	(700-1,000)
29. *Bab*	1954	28.3[E]	498[E]	5,502	6,000	(?)	(2,000)
32. *Umm Shaif***	1958	65.2	549	4,451	5,000	(?)	(1,850-2,300)
62. *Mubarras***	1969	7.7	15	1,985	2,000	(?)	(150-2,000)
64. *Abu al Bukhoosh***[a]	1966	21.9	29	471	500	(?)	(150-280)
236. *Saath al Raazboot***	1969	--	--	500	500	(low?)	
237. *Sahil*	1970	--	--	500	500	(low?)	
239. *Shah*	1966	--	--	500	500	(low?)	
Combination Giants							
261. *Arzanah*	1973	--	--	400	400	(?)	
Probable Giants							
*Umm Addalkh***	1968	--	--	300	300	(low?)	
Possible Giants							
*Al Bunduq***[a]	1965	5.0	5	95	100	(low?)	
*Ghasha***	1970	--	--				
*Nasr***	1971	--	--				
Rumaitha	1969	--	--				
Abu Dhabi Total		510.3	3,351	39,449	42,800	(?)	(25,000-43,000)
Post-1975 Discoveries							
Qusahwira	1976	--	--	700	700	(low?)	

[a]Combined with *Sassan*** (Iran) to have a total recovery of 2000 million barrels.

[b]Shared 50/50 with Qatar for a combined total of 200 million barrels.

NOTES: Discrepancies between published total reserve estimates for a country and for the sum of field reserve estimates in that country reach their zenith in Abu Dhabi. Estimates of Abu Dhabi reserves have consistently been two to three times higher than the sum of estimates of reserves in the individual fields of Abu Dhabi. The production plans, the water injection plans, reservoir volume parameters, and the conservative operating practices of the companies in Abu Dhabi are strong evidence that estimates for recovery in individual fields have been highly understated. The individual field estimates shown here may still underestimate ultimate recovery potential.

Zakum may have 40 to 50 billion barrels of oil-in-place. With additional development, some of which may be expensive by Middle East standards, total production capacity in the field could reach 1.7 MMb/d. Massive water injection projects have been recently completed in *Zakum, Bu Hasa, Asab, Bab,* and *Umm Shaif,* promising continuing high production capabilities for the next 15 to 30 years. *Bu Hasa* currently has a production capacity of 690 Mb/d, *Asab* 460 Mb/d, and *Bab* 130 Mb/d. *Bab* is relatively undeveloped compared with the other super-giants in Abu Dhabi because of its lower productivity per well and hence higher cost. Plans to develop *Mubarras* to 100-200 Mb/d have been indicated, but development has been slowed by high costs ($3 per barrel or more). *Saath al Raazboot, Sahil,* and *Shah* are indicated as minimal giants here, although each is probably larger. *Saath al Raazboot* has a planned sustained capacity of 80 Mb/d. Production in *Sahil* began in 1976 and is planned to begin in *Shah* in 1979. Development in both has been constrained by existing overcapacity in the other onshore fields.

The *Arzanah* discovery well-tested 4000 b/d and 90 MMcfd. Capacity for 40 Mb/d is currently being developed. but increases may be possible though expensive. *Umm Addalkh* is to be developed to a production capacity of 30-50 Mb/d. Production began in 1975 at *Al Bunduq* at a lower initial capacity than indicated originally (see also Qatar notes). By the end of 1976, three successful wells had been drilled in both *Ghasha* and *Nasr,* but no development plans have been announced. Extension wells have been drilled in *Rumaitha,* which appears to be a sizeable reservoir of relatively low permeability and hence high cost. The estimate for *Qusahwira* is an initial one that is likely to increase with further development.

Table A.50

GIANT OIL FIELDS AND RECOVERABLE OIL RESOURCES OF BAHREIN AS OF DECEMBER 31, 1975

(In millions of barrels)

Rank/Field	Discovery Year	1975 Production	Cumulative Production	Reserves	Total Recovery (With Range of Estimates)	
Giants						
123. *Awali*	1932	22.3	591	409	1,000	(650-1,050)
Bahrein Total		22.3	591	409	1,000	(850-1,050)

NOTES: *Awali* is the only field in Bahrein. Production is slowly declining and the field is likely to be nearly depleted by the year 2000.

Table A.51

GIANT OIL FIELDS AND RECOVERABLE OIL RESOURCES OF DUBAI AS OF DECEMBER 31, 1975

(In millions of barrels)

Rank/Field	Discovery Year	1975 Production	Cumulative Production	Reserves	Total Recovery (With Range of Estimates)	
Giants						
82. *Fateh***	1966	52.9	279	1,221	1,500	(1,000-1,600)
109. *Fateh SW***	1970	40.0	118	982	1,100	(550-1,100)
Probable Giants						
*Rashid***	1973	--	--			
Dubai Total		92.8	398	2,202+	2,600+	(2,300-2,700)

NOTES: The estimates for total reserves in Dubai and for the sum of reserves in *Fateh* and *Fateh SW* are in close agreement in the various sources. Large water injection facilities have just been installed in both fields, resulting in modest increases in production since 1975. *Rashid* is to be developed with production planned to begin by 1979. It is probably a combination giant as it has two gas-condensate zones as well as an oil zone.

Table A.52

GIANT OIL FIELDS AND RECOVERABLE OIL RESOURCES OF IRAN AS OF DECEMBER 31, 1975

(In millions of barrels)

Rank/Field	Discovery Year	1975 Production	Cumulative Production	Reserves	Total Recovery (With Range of Estimates)		
Giants							
6. *Ahwaz*	1958	374.3	1,810	15,690	17,500	(?)	(6,000-13,200)
8. *Marun*	1964	434.7	2,627	13,373	16,000	(?)	(6,000-12,000)
9. *Gach Saran*	1928	248.6	3,716	11,784	15,500	(?)	(8,000-12,000)
10. *Agha Jari*	1938	302.9	6,205	7,795	14,000	(?)	(8,000-14,000)
17. *Fereidoon*[**a]	1966	26.4	30	1,970	2,000	(?)	(1,300-6,000)
41. *Rag-e Safid*	1964	78.3	333	3,667	4,000	(?)	(2,500-4,500)
43. *Bibi Hakimeh*	1961	95.6	1,211	2,589	3,800	(?)	(2,500-9,000)
48. *Karanj*	1963	76.3	616	1,884	2,500	(low?)	(1,300-2,000)
49. *Paris*	1964	71.2	683	1,817	2,500	(?)	(1,800-3,000)
52. *Haft Kel*	1928	10.8	1,682	718	2,400	(?)	(1,800-3,000)
60. *Darius-Kharg*[*]	1961	42.4	479	1,521	2,000	(?)	(1,000-1,600)
64. *Sassan*[**b]	1966	64.3	414	1,086	1,500		(1,200-1,500)
76. *Ardeshir*[**]	1974	--	--	1,600	1,600	(?)	(1,600)
79. *Dehluran*[c]	1972(1971)	--	--	700	700	(low?)	
83. *Mansuri*	1963	4.7	9	1,491	1,500	(low?)	(1,500-2,000)
93. *Masjid-e-Suleiman*	1908	4.7	1,195	105	1,300	(low?)	(1,300-2,900)
128. *Kupal*	1964	5.2	23	977	1,000	(low?)	(1,000)
129. *Pazanan*	1937	11.0	151	849	1,000	(low?)	(1,000-3,500)
165. *Ab Teymur*	1968	--	--	750	750	(low?)	
223. *Binak*	1959	11.8	123	377	500	(?)	(500-1,000)
229. *Lab-e Safid*	1969	6.6	18	482	500	(?)	
232. *Naft-i-Safid*	1935	13.2	280	220	500	(?)	(500-1,000)
240. *Shurom*	1970	--	--	500	500	(?)	
Probable Giants							
Chesmeh Khush	1967	--	--	300	300	(low?)	
Cyrus[**]	1962	10.8	59	341	400	(?)	(500-800)
Kilar Karim	1968	--	--	300	300	(low?)	
Nowrouz[**]	1966	8.6	55	370	425	(?)	(410-1,000)
Rakhsh[**]	1969	10.9	63	237	300	(?)	(250-800)
Ramin	1966	--	--	250	250	(low?)	
Possible Giants							
Alborz	1956	--	5				
Bahregansar[**]	1961	6.2	103	97	200	(?)	(150-1,000)
Bushgan	1963	--	--	100	100	(low?)	(100-1,000)
Danan	1971	--	--	150	150	(low?)	
Esfandiar[**d]	1966	--	--				

Table A.52--continued

Rank/Field	Discovery Year	1975 Production	Cumulative Production	Reserves	Total Recovery (With Range of Estimates)	
Gulkhari	1963	--	--	200	200 (?)	
Kuh-i-Mund	1930	--	--			
Kuh-i-Rig	1967	--	--			
Lali	1938	--	59	91	150 (?)	
Maleh Kuh	1968	--	--	100	100	
Nargesi	1975	--	--			
Paydar	1974	--	--			
Qaleh Nar	1975	--	--	125	125 (low?)	
Ramshir	1961	5.8	32	168	200 (low?)	
*Rostam***	1966	7.3	82	118	200 (?)	(150-1,050)
Sarkan	1967	--	--	200	200 (?)	
Sarvestan	1973	--	--			
Shadegan	1968	--	--	150	150	
Siah Makan	1975	--	--			
Susangerd	1967	--	--			
Others		13.4	122	1,573	1,695	
Iran Total		1,946.0	22,185	76,815+	99,000+	(72,000-105,000)
Post-1975 Discoveries						
*Henjam***[e]	1976	--	--			

[a]Combined with *Marjan*** (Saudi Arabia) to have a total recovery of 10,000 million barrels.

[b]Combined with *Abu al Bukhoosh*** (Abu Dhabi) to have a total recovery of 2000 million barrels.

[c]Combined with *Abu Ghirab* (Iraq) to have a total recovery of 1500 million barrels.

[d]Combined with *Lulu*** (Neutral Zone); the field is a moderate-size structure for which no development plans have been announced.

[e]Combined with *Khasab*** (Oman).

NOTES: The data available for fields in Iran are strongly determined by economic considerations. In the fields for which the Oil Service Company of Iran (OSCO, formerly the Consortium) is the operator, production capacity has been developed according to the rational criterion of increasing marginal costs. In practice this has meant that almost no capacity with an investment cost of more than $200 per daily barrel has been developed. Many discoveries with healthy test flows by anything other than Middle East standards remain undeveloped. Moreover, most of the published reserve estimates for Iranian fields assume primary recovery only, which generally averages 20% in Iran. Given further exploration, development, and the application of secondary recovery to the approximately 375 billion barrels of original-oil-in-place in known Iranian fields, total recovery in Iran will easily exceed 100 billion barrels.

The reserve estimates used here include conservative estimates of secondary recovery for those fields where such operations have been installed or announced. A massive gas-injection program for *Ahwaz, Marun, Gach Saran, Agha Jari, Bibi Hakimeh, Karanj, Paris,* and *Haft Kel* is being initiated. Each of the four largest fields--*Ahwaz, Marun, Gach Saran,* and *Agha Jari*--have a production capacity of more than 1.0 MMb/d. A similar capacity was originally thought possible for *Karanj,* but was never realized. The original enthusiasm for *Bibi Hakimeh* (with an indicated potential up to 850 Mb/d) and *Paris* (with an indicated potential of 680 Mb/d) has diminished as well. It could not be determined whether the reduction in expectations for these three fields was the result of reduced geological potential or merely higher than anticipated costs of production. If considerable infill drilling occurs in these fields and they are produced for 50 to 75 more years, total recovery could prove to be 50% or more greater than the estimates given here.

The ultimate size of *Fereidoon-Marjan* (also Saudi Arabia) is yet to be evident, given its leisurely development to date. The estimate for *Rag-e Safid* could exclude the deeper, heavier (26° API) Sarvak (Bangestan) reservoir. The estimate shown here for *Darius-Kharg* assumes some water injection. Production is expected to reach 200 Mb/d in *Sassan* in 1978 and in *Ardeshir* in 1980. *Dehluran* could easily prove to be substantially larger than indicated here, but it probably contains heavy, high-sulfur oil. With secondary recovery, the estimate for the medium gravity (28° API) *Mansuri* field is likely to increase. *Masjid-e-Suleiman,* the first oil discovery in the Middle East, has nearly reached the limits of primary recovery. *Kupal* has been asserted to have a potential of 165 Mb/d. The oil reservoir at *Pazanan* underlies a super-giant gas cap with estimated reserves of 50 Tcf. Total recovery of petroleum liquids from the field could prove to be an amount several times greater

Table A.52--continued

than indicated here. The development of *Ab Teymur* has been constrained by relatively high costs (moreover, the oil is probably heavy). *Binak* is a small structure with excellent per-well productivity (over 20 Mb/d per well in 1977). *Lab-e Safid* has been asserted to have a potential of 110 Mb/d. *Naft-i-Safid* also has a gas cap of giant dimensions. *Shurom* is currently being developed. Because it is located in the Zagros Mountains, its development costs are quite high for an onshore field in the Middle East.

Chesmeh Khush is currently being developed. It has been asserted to have a potential of 125 Mb/d. Both *Cyrus* and *Nowrouz* produce heavy oil (18° and 21° API, respectively). Their offshore locations and the depths of the producing reservoirs may hamper recovery efforts. *Kilar Karim* and *Ramin* are relatively recent OSCO discoveries that may ultimately prove to be small giants. *Rakhsh* was once thought to have a 100 Mb/d potential. The planned water injection operations may boost production up close to that level.

There are a considerable number of undeveloped discoveries onshore in Iran. None appears to be a supergiant. However, most appear to be in structures with the area and thickness suitable for a giant field. Nearly all could be developed for capital and operating costs under $2 per barrel. *Alborz* had several highly productive wells (30 Mb/d each), but the high pressures of the reservoir appear to have precluded further development. *Bahregansar* has been asserted to be a giant, but that status appears to be an unlikely prospect for the first offshore discovery in Iran. *Bushgan, Danan, Nargesi, Paydar, Qaleh Nar, Shadegan,* and *Siah Makan* are recent OSCO discoveries in small- to medium-size structures. *Esfandiar* is still undeveloped, possibly because of high costs and an unsettled offshore boundary. *Gulkhari, Kuh-i-Mund,* and *Susangerd* are undeveloped heavy oil discoveries in the OSCO area. *Kuh-i-Rig* is a smaller field near *Shurom* covering several square miles. *Lali,* one of the earlier discoveries in the OSCO area, has a sizeable reservoir volume, but is plagued by low well productivity resulting from relatively poor porosity and permeability. *Maleh Kuh* and *Sarkan* are located in rugged terrain in the northern part of the OSCO area and thus have higher development and transport costs per barrel of production. *Ramshir* has a relatively thin oil column (250 feet in the Asmari) and oil with a high salt content. Nonetheless it could be a minor giant. *Rostam* may ultimately reach the giant status some sources have claimed for it. However, its production to date has not yet indicated a field of giant stature. Several successful extensions have been drilled already in *Sarvestan*. *Henjam-Khasab* is a large gas condensate discovery in the Straits of Hormuz.

Table A.53

GIANT OIL FIELDS AND RECOVERABLE OIL RESOURCES OF IRAQ AS OF DECEMBER 31, 1975

(In millions of barrels)

Rank/Field	Discovery Year	1975 Production	Cumulative Production	Reserves	Total Recovery (With Range of Estimates)		
Giants							
5. *Rumaila*	1953	355.0[E]	2,523[E]	17,477	20,000	(low?)	(18,000-21,000)
7. *Kirkuk*	1927	350.0[E]	6,933[E]	9,067	16,000	(?)	(15,000-16,500)
35. *Zubair*	1948	73.0[E]	776[E]	3,724	4,500	(?)	(1,900-4,600)
57. *Bai Hassan*	1953	14.0[E]	250[E]	1,750	2,000	(?)	(1,200)
59. *Buzurgan*	1969	--	--	2,000	2,000	(low?)	
79. *Abu Ghirab*[a]	1971	--	--	800	800	(low?)	
125. *Fauqi*[b]	1974	--	--	1,000	1,000	(low?)	
126. *Hamrin*	1973	--	--	1,000	1,000	(low?)	
127. *Jambur*	1954	2.0[E]	67[E]	933	1,000	(?)	
230. *Luhais*	1961	--	--	500	500	(?)	
233. *Nahr Umr*	1948	--	--	500	500	(?)	
Probable Giants							
Qayarah	1927	2.5[E]	36[E]				
Possible Giants							
Rachi	1957	--	--				
Ratawi	1950	--	--				
Siba	1968	--	--				
Others		6.5	309	155	464		
Iraq Total		803.0	10,894	38,906+	49,800+		(43,000-48,000)
Post-1975 Discoveries							
Majnoon	1976	--	--				(1,000-2,000)

[a]Combined with *Dehluran* (Iran) to have a combined recovery of 1500 million barrels.

[b]Extends into Iran, but so far only drilled in Iraq.

NOTES: Published reserve estimates for many of the Iraqi oil fields are likely to be conservative, partly because of the deliberate pace of development in Iraq before 1972 and partly because of the poor quality of the oil in many of the known fields. The estimates given here are minimum ones. There is a strong probability that there will be substantial reserve growth in known fields and that there will be additional giant fields discovered in Iraq. The Iraqi government has claimed that Iraq has reserves of 75 billion barrels, an estimate that is certainly not incredible for estimated total recovery.

The productive limits of *Rumaila* (including *North Rumaila*) are not yet fully determined; the field may extend into Kuwait. It could easily produce 2 MMb/d. *Kirkuk*, considered to be a model of efficient development, may also have more reserves than indicated here. Both *Rumaila* and *Zubair* have deeper zones containing heavier oil that are yet to be developed. Plans have been announced to expand the combined capacity of *Bai Hassan* and *Jambur* to as much as 300 Mb/d. *Buzurgan*, *Abu Ghirab*, and *Fauqi* are large structures in southeast Iraq with several producing zones containing heavy (23° API), high-sulfur (3.7%) crude oil with significant amounts of corrosive metals. The reported productive areas and oil columns of the three fields suggest significantly higher ultimately recoverable amounts than indicated here, particularly if their enhanced recovery

Table A.53--continued

potential is included. *Hamrin* is a large anticline; the discovery well tested 10 Mb/d. (Like *Bai Hassan* and *Jambur*, it was drilled in 1927-1930, at which time "shows" of oil and gas were found in it.) *Luhais* and *Nahr Umr* are being developed with initial capacities of 50 Mb/d and 40 Mb/d, respectively. Both are shown here as minimal giants, although they could be considerably larger.

Qayarah (including *Jawan*, *Najmah*, and *Qasab*) is a large accumulation of heavy (11°-19° API), high-sulfur oil that has been produced only for asphalt and residual fuel oil. The quality of the crude oil is not significantly different from that of producing giant fields in California and Venezuela. Production is simply too expensive by Middle East standards. Given its area and thickness, it could be a super-giant of the next century. *Rachi* and *Ratawi* are to be developed; no planned capacities have been announced. The *Siba* discovery well tested 6 Mb/d (reported as "oil shows" in the unique jargon of Middle East oil) in a 50-foot thick zone. The estimates for *Majnoon* were based on the discovery well only, which found a thick (1150-foot) oil column containing heavy, high-sulfur oil. An allegedly massive discovery west of Baghdad was announced in 1975; since then there has been no mention of it. In 1977, India's state-owned Oil and Natural Gas Commission abandoned as a "failure" the *Abu Khaimah* exploratory well testing 15 Mb/d.

SOURCE: H. V. Dunnington, "Generation, Migration, Accumulation, and Dissipation of Oil in Northern Iraq," in L. G. Weeks, ed., *Habitat of Oil*, American Association of Petroleum Geologists, Tulsa, Okla., 1958, pp. 1194-1251.

Table A.54

GIANT OIL FIELDS AND RECOVERABLE OIL RESOURCES OF ISRAEL AS OF DECEMBER 31, 1975

(In millions of barrels)

Rank/Field	Discovery Year	1975 Production	Cumulative Production	Reserves	Total Recovery (With Range of Estimates)
Others		0.3	15	1	16
Israel Total		0.3	15	1	16

NOTE: There are no oil fields of significance in Israel.

Table A.55

GIANT OIL FIELDS AND RECOVERABLE OIL RESOURCES OF KUWAIT AS OF DECEMBER 31, 1975

(In millions of barrels)

Rank/Field	Discovery Year	1975 Production	Cumulative Production	Reserves	Total Recovery (With Range of Estimates)		
Giants							
2. *Burgan*	1938	550.0[E]	14,840[E]	57,160	72,000	(?)	(66,000-72,000)
23. *Raudhatain*	1955	80.0[E]	1,587[E]	6,113	7,700	(?)	(7,700-9,000)
42. *Sabriyah*	1956	4.0[E]	208[E]	3,792	4,000	(?)	(4,000-7,000)
61. *Minagish*	1959	21.0[E]	204[E]	1,796	2,000	(?)	(2,000)
87. *Umm Gudair*[a]	1962	11.6[E]	124[E]	876	1,000	(?)	(1,000)
Possible Giants							
Bahrah	1956	--	--				
Khasman	1963	--	--				
Kuwait Total		666.6	16,963	69,737(+)	86,700(+)		(85,000-87,000)

[a]Combined with *S. Umm Gudair* (Neutral Zone) to have a combined recovery of 1500 million barrels.

NOTES: Because of the low levels of current and cumulative production relative to estimated reserves in Kuwait, estimates of total recovery are uncertain. Because production practices have been highly conservative, because development has been limited to only the least expensive production, and because reservoir characteristics are very good, the estimates shown here probably understate the ultimate amounts of crude oil which could be recovered from the known fields of Kuwait given extensive infill drilling and enhanced recovery operations. The reserve data shown assume pressure maintenance.

The reserve estimates for *Burgan* (including *Ahmadi* and *Magwa*), the second largest oil field in the world and the least expensive to produce, imply a recovery factor of 45%-50% in its Middle Cretaceous Wasia group sandstones. This recovery factor, though high in terms of average current recovery worldwide, may prove to be low given the high displacement efficiency likely to be achieved from waterflooding the highly porous and permeable *Burgan* sandstones, and from producing them at relatively low rates. *Burgan* also is asserted to have substantial amounts of heavy oil in Eocene and Upper Cretaceous reservoirs; these are not included in this compilation for lack of information about their producibility and economics. *Raudhatain* has a producing section 1500 feet thick. Both *Raudhatain* and *Sabriyah* have oil in several reservoirs. *Sabriyah* is likely to become a supergiant once it is subject to waterflooding. *Minagish* contains oil in three reservoirs, two containing heavy (20° API) oil of which production has yet to begin. Production from *Umm Gudair* is moderately heavy (27° API) and thus constrained by market factors. Two small fields, *Bahrah* and *Khasman*, have been discovered but not yet developed. Both appear to have relatively low productivity wells, indicating that they will be relatively expensive (by Middle East standards) to develop. *Rumaila* (Iraq) may extend into Kuwait. Several other discoveries not included here have been made in Kuwait, but have been suspended or abandoned because they were non-commercial (again, by presumably stringent Middle East standards).

SOURCE: M. Adasani, "The Greater Burgan Field," *Fifth Arab Petroleum Congress, Proceedings*, Vol. II, Cairo, 1963.

Table A.56

GIANT OIL FIELDS AND RECOVERABLE OIL RESOURCES OF THE NEUTRAL ZONE AS OF DECEMBER 31, 1975

(In millions of barrels)

Rank/Field	Discovery Year	1975 Production	Cumulative Production	Reserves	Total Recovery (With Range of Estimates)		
Giants							
4. *Khafji*** [a]	1960(1951)	117.0[E]	1,406	6,094	7,500	(?)	
33. *Wafra*	1953	43.3	905	4,095	5,000	(?)	(1,700-5,300)
87. *S. Umm Gudair* [b]	1966(1962)	18.1	134	366	500	(?)	(425-525)
Combination Giants							
266. *Dorra***	1967	--	--	340	340	(?)	(340)
Probable Giants							
*Hout***	1963	3.0[E]	39	161	200	(low?)	(200-1,000)
Possible Giants							
*Lulu*** [c]	1967	--	--	10	10		(10)
Others		0.9	13	17	30		
Neutral Zone Total		182.3	2,497	11,083	13,580		(7,000-19,600)

[a]Combined with *Safaniya*** (Saudi Arabia) to have a total recovery of 30,000 million barrels (24,000-30,000 million range).

[b]Combined with *Umm Gudair* (Kuwait) to have a total recovery of 1500 million barrels.

[c]Combined with *Esfandiar*** (Iran); the field is a moderate-size structure for which no development plans have been announced.

NOTES: *Safaniya-Khafji* is considered to be the largest offshore field in the world. During the past several years, reserve estimates for *Wafra*, and thus for the Neutral Zone as a whole, have varied considerably, primarily because it contains heavy (18°-24° API), high-sulfur (4.7%) oil with limited marketability. Because there are very large amounts of oil-in-place, it is listed here as a minimal super-giant. *S. Umm Gudair* is the southern extension of *Umm Gudair*. Both *Dorra* and *Hout* may prove to be considerably larger than indicated here, some sources indicating that they occupy sizeable structures. *Dorra* is currently undeveloped and may be both a giant oil field and a giant gas field (in different producing zones). *Lulu* is a small extension of *Esfandiar*; the estimates here are estimates of recovery from only two wells drilled by the end of 1975.

Table A.57

GIANT OIL FIELDS AND RECOVERABLE OIL RESOURCES OF OMAN AS OF DECEMBER 31, 1975

(In millions of barrels)

Rank/Field	Discovery Year	1975 Production	Cumulative Production	Reserves	Total Recovery (With Range of Estimates)		
Giants							
142. *Fahud*	1964	37.9	453	447	900	(?)	(1,000-1,175)
199. *Yibal*	1962	34.5	149	451	600	(?)	(650-675)
Probable Giants							
Natih	1963	17.7	212	188	400	(?)	(600-650)
Others		34.5	66	1,234	1,300	(?)	
Oman Total		124.6	880	2,320	3,200		(4,000-7,000)
Post-1975 Discoveries							
*Khasab*** [a]	1977(1976)	--	--				

[a]Combined with *Henjam*** (Iran).

NOTES: Published estimates of reserves for the three giant fields of Oman appear to be overstated, given the declining production in each and the prospects of continued declines. All three fields have had active water-injection operations for several years. *Natih,* if it proves to be a giant, will most likely be a combination giant. The fields discovered in Oman from 1970 to 1973 (*Alam Shua, Huwaisa, North Ghaba, Sai Rawal,* and *Saih-Nihaydah*) appear to be in the sub-giant category (100-500 million barrels each), including their oil-equivalent gas resources. *Henjam-Khasab* is a gas-condensate discovery (6000 b/d) on a large offshore structure in the Straits of Hormuz.

Table A.58

GIANT OIL FIELDS AND RECOVERABLE OIL RESOURCES OF QATAR AS OF DECEMBER 31, 1975

(In millions of barrels)

Rank/Field	Discovery Year	1975 Production	Cumulative Production	Reserves	Total Recovery (With Range of Estimates)		
Giants							
44. *Dukhan*	1940	59.5	1,561	1,289	2,850		(2,400-2,900)
55. *Idd el Shargi***	1960	5.0	150	1,950	2,100	(high?)	(2,100)
95. *Bul Hanine***	1970	50.7	172	1,078	1,250	(low?)	(120-1,000)
105. *Maydan-Mahzan***	1963	39.2	439	761	1,200	(low?)	(1,100-10,000)
Possible Giants							
*Al Bunduq*** [a]	1965	5.0	5	95	100	(low?)	
Qatar Total		159.5	2,327	5,173	7,500		(7,600-8,100)

[a]Shared 50/50 with Abu Dhabi for a combined total of 200 million barrels.

NOTES: All the fields of Qatar are subject to production restrictions. In 1975, production was further constrained by international demand. Waterflooding has been going on in *Dukhan* for some years and has begun in *Maydan-Mahzan.* Although the latter could be larger than indicated here, the size of the structure makes it unlikely that it could ever exceed 2000 million barrels. The published estimates for *Idd el Shargi* are questionable. It appears to have sufficient reservoir volume for 2 billion barrels' ultimate recovery, but generally low permeability and hence low productivity per well and relatively high costs per barrel of production may limit recovery. Given the rapid development of *Bul Hanine,* published estimates of its potential appear to be low. *Al Bunduq* began production in 1975, four years after initially planned. Production capacities as high as 50 Mb/d have been mentioned for it; however, its relatively high development costs ($2000 per daily barrel of capacity) have limited initial capacity to 20 Mb/d. At best, it is a minimal giant.

Table A.59

GIANT OIL FIELDS AND RECOVERABLE OIL RESOURCES OF SAUDI ARABIA AS OF DECEMBER 31, 1975

(In millions of barrels)

Rank/Field	Discovery Year	1975 Production	Cumulative Production	Reserves	Total Recovery (With Range of Estimates)		
Giants							
1. *Ghawar*	1948	1,665.6	12,542	70,458	83,000	(low?)	(60,000-83,000)
4. *Safaniya*[**a]	1951	237.7	3,264	19,236	22,500	(low?)	(18,000-25,000)
12. *Abqaiq*	1940	281.3	5,058	7,442	12,500	(low?)	(8,400-12,500)
14. *Berri*[*]	1964	153.8	885	11,115	12,000	(low?)	(5,900-7,950)
16. *Manifa*[**]	1957	8.2	131	10,869	11,000	(low?)	(1,000-11,000)
17. *Marjan*[**b]	1967(1966)	2.7	8	7,992	8,000	(low?)	(700-24,000)
20. *Qatif*	1945	23.2	474	8,526	9,000	(low?)	(3,150-9,100)
21. *Khurais*	1957	13.1	64	8,436	8,500	(low?)	(4,350-7,650)
22. *Zuluf*[**]	1965	24.3	152	8,348	8,500	(low?)	(3,200-5,400)
26. *Shaybah*[c]	1968	--	--	7,000	7,000	(low?)	(1,000-4,000)
27. *Abu Sa'fah*[**d]	1963	21.1	270	6,330	6,600		(3,250-6,600)
39. *Khursaniyah*	1956	13.9	486	3,514	4,000		(2,000-6,500)
94. *Rimthan*	1974	--	--	1,300	1,300	(low?)	(300-1,500)
114. *Abu Hadriya*	1940	20.1	318	737	1,055	(low?)	(1,000-1,055)
117. *Dammam*	1938	6.7	545	500	1,045		(650-5,500)
120. *Harmaliyah*	1971	15.6	75	950	1,025	(low?)	(50-1,025)
136. *Fadhili*	1949	5.4	162	798	960	(?)	(550-1,000)
227. *Jana*[**]	1967	--	--	500	500	(low?)	(40-550)
231. *Maharah*[**]	1973	--	--	500	500	(low?)	(70)
Probable Giants							
Abu Jifan	1973	--	--	279	279	(low?)	(280)
Lawhah[**]	1975	--	--	276	276	(low?)	(275)
Mazalij	1971	--	--	338	338	(low?)	(340-1,000)
Possible Giants							
Dibdibah	1975	--	--	7	7	(low?)	
El Haba	1973	--	--	57	57	(low?)	
Juraybi'at	1968	--	--	2	2	(low?)	
Karan[**]	1967	--	--	10	10	(low?)	
Qirdi	1973	--	--	36	36	(low?)	(35-1,000)
Ramlah	1974	--	--	15	15	(low?)	
Ribyan[**]	1975	--	--	9	9	(low?)	
Watban	1975	--	--				
Others		--	--	186	186		
Saudi Arabia Total		2,492.7	24,435	175,765	200,200		

[a]Combined with *Khafji*[**] (Neutral Zone) to have a total recovery of 30,000 million barrels (24,000-30,000 million barrel range).

[b]Combined with *Fereidoon*[**] (Iran) to have a total recovery of 10,000 million barrels.

[c]Includes *Zarrara* (Abu Dhabi): resolution of the border dispute in 1975 gave the entire structure to Saudi Arabia.

[d]Revenues from production are split with Bahrein.

Table A.59--continued

NOTES: Saudi Arabia not only has more known recoverable oil resources than any other country in the world, but it also has the most rapidly growing known oil resources of any country. From 1970 to 1976, average proved and probable gross reserve additions were more than 11 billion barrels per year. Annual gross reserve additions in Saudi Arabia should continue to be substantial for several more years, if not decades. Several new fields are being discovered every year, although the average size of discovery has declined significantly. Most of the published reserve estimates for Saudi fields assume only 15%-30% recovery of oil-in-place, a very low figure for ultimate recovery of medium-to-high-gravity oil from highly porous and permeable reservoirs (which most of the Saudi fields are). Pressure maintenance facilities have been installed in only a few fields (gas injection in *Abqaiq* and *Ghawar*; water injection in *Abqaiq, Abu Hadriya, Berri, Ghawar, Harmaliyah,* and *Khursaniyah*). Well densities are generally extremely low. Given the immense amounts of known oil-in-place (most likely between 700 and 800 billion barrels), the potential for secondary and enhanced recovery operations, a potential that is unlikely to be fully realized until the next century, is huge.

Because Saudi reserves have been increasing substantially from year to year, especially from extensions and revisions in known fields, any estimate of known reserves by field is likely to understate the potential of individual fields. For lack of more definitive information, the reserve estimate for Saudi Arabia used here is Aramco's estimate of proved and probable reserves as of the end of 1975. Estimates of Saudi proved reserves by field as of the end of 1975 have been published (Hasson, Mason, and Moore). However, probable reserves (defined as oil known to be in place and estimated to be recoverable with further drilling or the installation of secondary recovery facilities) are another 68 billion barrels. The estimates presented here for known giants in Saudi Arabia assume that estimates from other sources that are higher than the published estimates of proved reserves are reasonable indicators of the proved and probable reserves as defined by Aramco. Because most of the fields are not fully developed and are still in the early stages of their productive life, there is still considerable potential for reserve growth in the Saudi fields. (Aramco has basically developed production capacity in line with marginal costs, concentrating initially on the lowest cost onshore fields.)

Ghawar is the largest oil field in the world, with a current production capacity exceeded by the total production of only three other countries and a potential capacity of 9.0 MMb/d. It covers more than 700,000 acres and currently has a well density of only one well per every two square miles. It would not be surprising if it ultimately produces more than 100 billion barrels. Because of their offshore location and hence more expensive production, *Safaniya, Manifa, Marjan, Zuluf,* and *Abu Sa'fah* have not been extensively developed, despite their super-giant status. All but *Abu Sa'fah* have a potential for substantial reserve growth, particularly *Safaniya* and *Zuluf*. *Safaniya* and *Zuluf* have the potential to produce 2.0 and 1.5 MMb/d, respectively. Both *Abqaiq* and *Berri* have been producing recently at a level around 800 Mb/d. *Qatif* and *Khursaniyah* are still largely undeveloped, despite their favorable location near Ras Tanura. When *Abu Jifan, Mazalij,* and *Qirdi* were discovered, there was some speculation that they would link up with *Khurais,* forming another *Ghawar*. The possibility is still in the realm of speculation. *Shaybah,* with an indicated potential of 500 Mb/d, is the only major discovery made to date by Aramco in the Rub'al Khali. Because of its isolated location and its relatively low indicated per-well production, it has the highest investment costs per daily barrel of capacity ($800 or more) of any field currently producing or under development in Saudi Arabia. *Rimthan* is included as a giant here on the strength of announced plans for a 150 Mb/d start-up in 1978. Production in both *Abu Hadriya* and *Harmaliyah* has reached 130 Mb/d in recent years, suggesting that their reserves are likely to be understated. *Jana* and *Maharah* are included here as minimal giants because various sources have indicated that they are among the more significant recent discoveries in Saudi Arabia.

Abu Jifan and *Mazalij* obviously extend beyond the boundaries of Aramco's Area 1 concession. The reserve estimate for *Lawhah* is based on just the initial two exploratory wells. All of the other recent discoveries are considered to be small or medium by Saudi standards. However, because it takes less than 1% of a *Ghawar* to constitute a giant field, many may prove to be giants once they are developed. The reserve estimates shown for all of the probable and possible giants are strictly estimates of proved reserves and appear in most cases to be estimates of recoverable oil from the discovery well only. *Barqan,* a 1969 Red Sea discovery, was asserted as a giant by Halbouty, et al. Because it was a gas-condensate discovery, it should be considered as a giant gas field if it proves to be a giant.

SOURCES:

1. *Aramco: A Review of Operations* (annual).
2. R. C. Hasson, J. F. Mason, and Q. M. Moore, "Petroleum Developments in Middle East Countries in 1975," *American Association of Petroleum Geologists Bulletin,* Vol. 60, No. 10, October 1976, pp. 1865-1899.
3. R. W. Powers, et al., *Geology of the Arabian Peninsula: Sedimentary Geology of Saudi Arabia,* Geological Survey Professional Paper 560-D, U.S. Government Printing Office, Washington, D.C., 1966.

Table A.60

GIANT OIL FIELDS AND RECOVERABLE OIL RESOURCES OF SHARJAH AS OF DECEMBER 31, 1975

(In millions of barrels)

Rank/Field	Discovery Year	1975 Production	Cumulative Production	Reserves	Total Recovery (With Range of Estimates)		
Others		13.9	22	178	200	(?)	(75-1,500)
Sharjah Total		13.9	22	178	200	(?)	(75-1,500)

NOTES: *Mubarek* is the only field discovered to date in Sharjah. Estimates of its potential have varied widely. Although the platform was designed for 100 Mb/d plus, production has yet to exceed 40 Mb/d. Because increased production is no longer anticipated, it does not merit giant status.

Table A.61

GIANT OIL FIELDS AND RECOVERABLE OIL RESOURCES OF SYRIA AS OF DECEMBER 31, 1975

(In millions of barrels)

Rank/Field	Discovery Year	1975 Production	Cumulative Production	Reserves	Total Recovery (With Range of Estimates)		
Giants							
86. *Souedie*	1959	49.3^E	265^E	1,235	1,500	(?)	(650 and up)
143. *Karatchouk*	1956	9.1^E	41^E	859	900	(?)	(900-2,500)
Possible Giants							
Kirbah	1962	--	--				
Tel Rumailan	1961	3.6^E	18^E	182	200	(?)	(200-1,500)
Others		1.9^E	2^E	198	200	(?)	
Syria Total		63.9^E	326^E	2,474+	2,800+	(?)	(1,700-7,300)

NOTES: Reliable data on Syrian oil production and reserves are not available. Estimates of both appearing in standard sources vary widely. The estimates of 1975 and cumulative production given here are probably accurate to a range of ± 10%. As the range of estimates indicates, the estimates of reserves are subject to considerably greater uncertainty.

Souedie and *Karatchouk* are unquestionably giant fields. The question is how big they are. Both have sufficient reservoir volume to hold several billion barrels of oil-in-place. Yet the oil of both is heavy (20°-25° API), and per-well productivity is low. Marketability is also limited by a high-sulfur content (3.5% and 4.5%, respectively). Both could be somewhat larger than shown here, assuming the initiation of expensive thermal recovery operations.

Kirbah (also known as *Khurbet*) has been indicated as a giant in some sources and as a small noncommercial discovery in others. *Tel Rumailan* has been asserted in some recent sources to be a giant and in others to be the least important producing field in Syria. In both cases, the assessment of potential could reflect different assumptions about viable economic production; both fields appear to have sufficient reservoir volume to be giants, given 20% to 30% recovery, yet both have heavy oil and low well productivity.

SOURCES: J. F. Mason, Q. M. Moore, and R. C. Hasson, "Petroleum Developments in Middle East Countries in 1972," *American Association of Petroleum Geologists Bulletin*, Vol. 57, No. 10, October 1973, pp. 2057-2084.

Table A.62

GIANT OIL FIELDS AND RECOVERABLE OIL RESOURCES OF TURKEY AS OF DECEMBER 31, 1975

(In millions of barrels)

Rank/Field	Discovery Year	1975 Production	Cumulative Production	Reserves	Tocal Recovery (With Range of Estimates)	
Others		22.2	273	277	550	
Turkey Total		22.2	273	277	550	(375-650)

NOTE: The largest oil fields discovered to date in Turkey are smaller than 100 million barrels' ultimate recovery.

Table A.63

KNOWN RECOVERABLE CRUDE OIL RESOURCES OF ASIA/OCEANIA BY FIELD CLASSIFICATION AS OF DECEMBER 31, 1975

(In millions of barrels)

Field Classification	Number of Fields	1975 Production	Cumulative Production	Reserves	Total Recovery
Giant fields	17	705.9	6,462	15,788	22,250
Combination giants	--	--	--	--	--
Probable giants	--	--	--	--	--
Possible giants[a]	20(+)	323.4	2,073	10,317(+)	12,390(+)
Other fields[a]		305.5	4,163	11,997	16,160
Total		1,334.8	12,698	38,102(+)	50,800(+)

[a]In order to have a complete allocation of estimated reserves, 60% of Chinese reserves that are not in known giant fields were allocated to possible giants; the remaining 40% are included in other fields.

NOTES: Because of the lack of information about most Chinese oil fields, a high proportion of the known recoverable oil resources of Asia/Oceania are estimated to be in potential giant fields. The 17 known giant fields contain 22.25 billion barrels, 43.8% of the known recoverable resource. There are at least 20 potential giant fields with 12.39 billion barrels (24.4%). Other fields have 16.16 billion barrels (31.8%). Of the 10.88 billion barrels in other fields outside of China, approximately 7.30 billion are in 47 large fields. Probably at least half of the estimated amount in other fields in China is in large fields as well, although the available data are inadequate for any definitive judgment here.

The known giants have 15.79 billion barrels of reserves (41.4% of the total). Potential giants have an estimated 10.32 billion barrels (27.1%), and the non-giant fields have 12.00 billion barrels (31.5%).

The quality of the information on known oil resources in Asia/Oceania varies considerably from country to country. For some, such as Australia, the data are well documented and quite reliable. There are still some significant uncertainties in those countries with recent discoveries such as India, Indonesia, and Malaysia. Less information is available about China than about any other country with several known giant fields.

SOURCE: W. R. Scheidecker, "Petroleum Developments in Far East in 1975," *American Association of Petroleum Geologists Bulletin,* Vol. 60, No. 10, October 1976, pp. 1908-1946 (also previous years).

Table A.64

GIANT OIL FIELDS AND RECOVERABLE OIL RESOURCES OF AUSTRALIA AS OF DECEMBER 31, 1975

(In millions of barrels)

Rank/Field	Discovery Year	1975 Production	Cumulative Production	Reserves	Total Recovery (With Range of Estimates)		
Giants							
96. *Kingfish***	1967	74.2	314	936	1,250	(low?)	(950-1,060)
149. *Halibut***	1967	46.6	304	546	850	(low?)	(440-650)
Possible Giants							
*Mackerel***	1969	--	--	300	300		(300)
Mereenie	1964	--	--	250	250		(200-300)
Others		15.1	150	500	650		
Australia Total		135.9	768	2,532	3,300		(2,300-3,100)

NOTES: Subject to some reserve growth, published estimates for oil reserves by field in Australia are relatively reliable. The estimates given here reflect an increase of approximately 400 million barrels in the Bass Strait fields, primarily *Kingfish* and *Halibut*, announced in October 1977. Given the announced intentions of Esso Australia to maintain production in these two fields at present rates (235 Mb/d and 140 Mb/d, respectively) for another five years, these estimates could be slightly conservative. Both fields are undergoing pressure maintenance. *Mackerel*, with a planned production of 90 Mb/d, could prove to be a minimal giant. *Mereenie* may be a combination giant. *Marlin* and *Snapper*, two gas-condensate offshore fields in the Gippsland Basin, are combination gas giants. The giant gas fields of the Northwest Shelf may contain as much as a billion barrels of natural gas liquids (not included in this tabulation). *Barrow Island*, with ultimately recoverable oil of 250 million barrels, is the only other sizeable oil field discovered to date in Australia.

SOURCES:

1. Australia and New Zealand Banking Group Limited, *Oil and Gas in Australia*, Melbourne, 1976.
2. L. R. Beddoes, Jr., ed., *Oil and Gas Fields of Australia, Papua New Guinea, and New Zealand*, Eurasia Press, Singapore, 1973.
3. E. F. Durkee, "Petroleum Developments in Australia in 1975," *American Association of Petroleum Geologists Bulletin*, Vol. 60, No. 10, October 1976, pp. 1957-1964.

Table A.65

GIANT OIL FIELDS AND RECOVERABLE OIL RESOURCES OF BRUNEI AS OF DECEMBER 31, 1975

(In millions of barrels)

Rank/Field	Discovery Year	1975 Production	Cumulative Production	Reserves	Total Recovery (With Range of Estimates)	
Giants						
116. *Southwest Ampa***	1963	33.5	298	752	1,050	(1,000-1,050)
121. *Seria**	1926	11.9	842	183	1,025	(1,025-1,730)
Possible Giants						
*Champion***	1970	12.2	40	285	325 (low?)	(150-200)
Others		8.3	34	266	300	
Brunei Total		65.9	1,214	1,486	2,700	(2,900-3,200)

NOTES: *Southwest Ampa* is also a large gas field with an estimated 1.4 Tcf of reserves in separate producing zones. *Seria*, one of the oldest giants in south and east Asia, is largely depleted; however, there is some potential for reserve growth from the application of enhanced recovery in its upper, heavy oil reservoirs. *Champion*, also a heavy oil field (23° API), is still being developed and could prove to be a giant with enhanced recovery.

SOURCE: H. B. Schaub and A. Jackson, "The Northwestern Oil Basin of Borneo," in L. G. Weeks, ed., *Habitat of Oil*, American Association of Petroleum Geologists, Tulsa, Okla., 1958, pp. 1330-1336.

Table A.66

GIANT OIL FIELDS AND RECOVERABLE OIL RESOURCES OF BURMA AS OF DECEMBER 31, 1975

(In millions of barrels)

Rank/Field	Discovery Year	1975 Production	Cumulative Production	Reserves	Total Recovery (With Range of Estimates)	
Others		7.8	379	71	450	
Burma Total		7.8	379	71	450	(450)

NOTES: Oil production in Burma began in 1887. Most of the production to date has come from two old, moderate-size fields, *Yanangyuang* and *Chauch*.

SOURCE: H. R. Tainsh, "Burma," in V. C. Illing, ed., *The Science of Petroleum: The World's Oil Fields--Eastern Hemisphere*, Oxford University Press, London, 1953.

Table A.67

GIANT OIL FIELDS AND RECOVERABLE OIL RESOURCES OF CHINA AS OF DECEMBER 31, 1975

(In millions of barrels)

Rank/Field	Discovery Year	1975 Production	Cumulative Production	Reserves	Total Recovery (With Range of Estimates)		
Giants							
30. *Ta-ch'ing*	1959	285.0[E]	1,725[E]	4,275	6,000	(?)	(3,000-6,000)
110. *Leng-hu*	1958	3.0[E]	70[E]	1,030	1,100	(?)	(1,100)
170. *Karamai*	1955	7.8[E]	60[E]	670	730	(?)	(730)
194. *Lung-nu*	1956	2.0[E]	40[E]	560	600	(?)	(600)
215. *Ya-erh-hsia*	1957	4.0[E]	60[E]	465	525	(?)	(525+)
Possible Giants							
Ch'ien-chiang[a]	1967 (?)	24.0[E]	120[E]				(to 4,200)
Fu-yu[a]	1960	19.0[E]	100[E]				
I-tu[a]	1970	22.0[E]	50[E]				
Lao-chun-mia	1939	2.0[E]	25[E]				(to 2,000)
P'an-shan[a]	1964 (?)	24.0[E]	70[E]				
Shengli[a]	1962	105.0[E]	465[E]				(to 6,000)
Ta-kang[*a]	1965	32.0[E]	125[E]				
Ta-king	1973 (?)	5.0[E]	10[E]				
Others		7.2[E]	80[E]				
China Total		542.0[E]	3,000[E]	20,000+	23,000+		(20,000-40,000)

[a]Area comprising several distinct fields.

NOTES: As Connell has expressed it most succinctly, "Fragments of information and infrequent issuance of questionable statistics lend but little authenticity or substance to any report on the petroleum industry within the People's Republic of China." The problems are particularly acute for any compilation of giant fields because what little data there are about "fields" often refer not to individual fields as defined in this report and elsewhere but to producing areas containing several fields. No data are available about the individual fields within these areas. There are also no estimates of reserves by field available except those based on Russian information that is more than fifteen years old. The Chinese are also slow to release information about major new discoveries. Given these overwhelming data limitations, any compilation of giant oil fields in China such as this one should be considered as highly tentative.

The rate of increase in production at *Ta-ch'ing* (Sung-liao Basin) has slowed considerably, indicating that it may be reaching a production plateau. Given its cumulative and current production, it appears to be at least a nominal super-giant. The recovery estimates for *Leng-hu* (Tsaidam Basin), *Karamai* aka *K'o-la-ma-i* (Dzungarian Basin), *Lung-nu* (Szechwan), and *Ya-erh-hsia* (Pre-Nan-Shan Basin) were derived by Meyerhoff from Russian sources. Because they are all in relatively remote areas, none has been extensively developed. Some sources have suggested that *Karamai* and *Ya-erh-hsia* could be larger than indicated here.

Little is known about the *Ch'ien-chiang* fields (Hupeh Province). Japanese sources have described the area as having 4200 million barrels of reserves in at least four separate fields. One of the several fields in the *Fu-yu* area (Sung-liao Basin) may be of giant size. Extraction is reported to be difficult. The fields of *I-tu* are often considered to be part of the *Shengli* area (North China Basin). There could be several separate giant fields in this heavily faulted area, which is asserted to have a potential equal to that of *Ta-ch'ing*. The area is still being developed. Akimov has asserted that *Lao-chun-mia* (Pre-Nan-Shan Basin) has reserves of 2000 million barrels, an estimate that could not be confirmed from other sources. The *P'an-shan* area (North China Basin) has several producing fields now being developed, the largest of which may prove to be a giant. There are several scattered fields in the *Ta-kang* area (North China Basin). Additional development could confirm several giants in the area. *Ta-king* (Sung-liao Basin) is a recently discovered field 30 to 60 miles from *Ta-ch'ing*. Several sources have touted it highly, and production goals have been mentioned which, if realized, will qualify it as a giant.

Given further exploration and development, it is highly probable that more giant fields will be discovered

Table A.67--continued

in China. Known discoveries that could grow into giants include *Nan-shan* (Tsaidam Basin), *Urho* (Dzungarian Basin), and *Yin-shan* (Szechwan).

The estimate of total recovery for China used here is a reasonable approximation of proved and probable reserves. Much higher estimates have been published. However, they appear to be estimates of total resource potential, a large proportion of which are only possible reserves at present.

SOURCES:

1. V. I. Akimov, "The Fuel and Power Base of the PRC," *Problems of the Far East*, No. 1, Joint Publications Research Service, No. 61955, May 19, 1974 (translated from the Russian original).

2. Central Intelligence Agency, *China: Oil Production Prospects*, Washington, D.C., June 1977.

3. H. R. Connell, "China's Petroleum Industry--An Enigma," *American Association of Petroleum Geologists Bulletin*, Vol. 58, No. 10, October 1974, pp. 2157-2172.

4. R. W. Hardy, *Chinese Oil: Development Prospects and Potential Impact*, Center for Strategic and International Studies, Washington, D.C., November 1976.

5. S. S. Harrison, *China, Oil, and Asia: Conflict Ahead?*, Columbia University Press, New York, 1977.

6. A. A. Meyerhoff, "Developments in Mainland China," *American Association of Petroleum Geologists Bulletin*, Vol. 54, No. 10, August 1970, pp. 1567-1580.

7. K. P. Wang, *The People's Republic of China: A New Industrial Power with a Strong Mineral Base*, U.S. Department of the Interior, Washington, D.C., 1975.

8. B. A. Williams, "The Chinese Petroleum Industry: Growth and Prospects," in *China: A Reassessment of the Economy*, a compendium of papers submitted to the Joint Economic Committee, Washington, D.C., July 1975, pp. 225-263.

Table A.68

GIANT OIL FIELDS AND RECOVERABLE OIL RESOURCES OF INDIA AS OF DECEMBER 31, 1975

(In millions of barrels)

Rank/Field	Discovery Year	1975 Production	Cumulative Production	Reserves	Total Recovery (With Range of Estimates)		
Giants							
80. *Bombay High***	1974	--	--	1,500	1,500	(?)	(500-2,000)
207. *Bassein North***	1975	--	--	550	550	(?)	(500-550)
Others		61.5	604	946	1,550		
India Total		61.5	604	2,996	3,600		(3,600)

NOTES: Extensive development is proceeding on both *Bombay High* and *Bassein North*. The reserve estimates shown here are those of the Oil and Natural Gas Commission of India, and may be subject to substantial future revision. By 1981, production is planned to be 200 Mb/d in *Bombay High* and 40 Mb/d in *Bassein North*.

Table A.69

GIANT OIL FIELDS AND RECOVERABLE OIL RESOURCES OF INDONESIA AS OF DECEMBER 31, 1975

(In millions of barrels)

Rank/Field	Discovery Year	1975 Production	Cumulative Production	Reserves	Total Recovery (With Range of Estimates)		
Giants							
40. *Minas*	1944	136.9	1,946	2,054	4,000	(?)	(4,000-7,200)
153. *Handil**	1974	1.6	2	798	800	(?)	
188. *Attaka***	1970	36.2	101	499	600	(?)	(350-400)
191. *Duri*	1941	10.4	258	342	600	(?)	(to 2,000)
202. *Bekasap*	1955	21.2	293	277	570	(?)	(570-600)
221. *Bangko*	1970	31.6	149	351	500	(?)	(300-700)
Possible Giants							
Jatibarang	1970	8.0	27	223	250	(?)	
Pematang	1959	15.1	146	134	280	(?)	(280-300)
Petani	1964	19.4	177	183	360	(?)	(280-360)
Rantau	1929	7.8	202	88	290	(?)	(250-290)
Sanga-Sanga	1897	1.6	245	25	270	(?)	(250-500)
Talang Akar	1922	0.3	241	9	250	(?)	(600-725)
Walio	1973	11.5	12	388	400	(?)	
Others		175.6	2,469	3,861	6,330		
Indonesia Total		477.2	6,267	9,233	15,500		(17,000-20,000)

NOTES: Unless recent published reserve estimates for Indonesia assume extensive secondary and enhanced recovery operations that are not now under way and that in many cases are not economic for field operators under current agreements, they are clearly overstated. The most recent estimates are beginning to reflect a greater realism, scaling down earlier, more optimistic estimates as production declines in all but the newest fields become apparent.

Minas is the largest field in southeast Asia and Oceania, but it does not appear to be a super-giant. The estimate here assumes roughly 55% recovery of the estimated 7 billion barrels of oil-in-place as a result of current water-injection operations. No reserve estimates have been published for *Handil*, but its production reached 200 Mb/d in early 1977, indicating that it is clearly a giant. *Attaka*, with production for five straight years at 100-120 Mb/d and a recently lengthened cost recovery period for the operators, is at least a marginal giant. The estimate for *Duri*, a heavy oil field with less than 10% primary recovery but excellent porosity and permeability, is based on extensive enhanced recovery operations (steam-soak and steam-flood). Estimates that are much higher are not credible given production to date. *Bekasap* and *Bangko*, two central Sumatran fields with declining primary production, are likely marginal giants, but both will require secondary recovery to become so.

Jatibarang was asserted to have a potential of 150 Mb/d; most of the wells drilled in this recent Pertamina discovery are still shut-in. *Pematang*, *Petani*, and *Rantau*, particularly *Petani*, could become giants or combination giants with secondary recovery. *Sanga-Sanga*, one of the oldest fields in Indonesia, has been undergoing additional development and may eventually reach the marginal giant status asserted by some. *Talang Akar* has been mentioned as a giant, but at best is a highly marginal candidate for that status. *Walio* has been asserted to be a giant with a production potential of 100-150 Mb/d, but the most recent evidence indicates that it may fall short of giant status. There are a number of smaller recent discoveries that are now being developed or are awaiting development.

SOURCE: R. R. Vincelette and A. R. Soeparjadi, "Oil-Bearing Reefs in Salawati Basin of Irian Jaya, Indonesia," *American Association of Petroleum Geologists Bulletin*, Vol. 60, No. 9, September 1976, pp. 1448-1462.

Table A.70

GIANT OIL FIELDS AND RECOVERABLE OIL RESOURCES OF JAPAN AS OF DECEMBER 31, 1975

(In millions of barrels)

Rank/Field	Discovery Year	1975 Production	Cumulative Production	Reserves	Total Recovery (With Range of Estimates)	
Others		4.4	132	43	175	
Japan Total		4.4	132	43	175	(165-200)

NOTE: Only several small fields have been discovered on and offshore Japan despite a relatively intensive search.

Table A.71

GIANT OIL FIELDS AND RECOVERABLE OIL RESOURCES OF MALAYSIA AS OF DECEMBER 31, 1975

(In millions of Barrels)

Rank/Field	Discovery Year	1975 Production	Cumulative Production	Reserves	Total Recovery (With Range of Estimates)		
Possible Giants							
*Baronia***	1970	12.3	16	334	350	(?)	
*Samarang***	1972	2.2	2	298	300	(?)	
Others		20.4	228	922	1,150		
Malaysia Total		34.9	246	1,554	1,800	(?)	(1,100-2,800)

NOTES: Development of the recently discovered offshore fields in Malaysia has been constrained because of disputes between the Malaysian government and the operating companies over possible nationalization. Since 1974, when the dispute began, information about their potential has been noticeably absent, particularly when compared with the optimistic projections immediately following discovery. With the disputes now being resolved, production plans suggest a more pessimistic outlook than before. *Baronia* reached 53 Mb/d and *Samarang* reached 62 Mb/d in 1977. Both are still being developed and could be marginal giants or combination giants.

Table A.72

GIANT OIL FIELDS AND RECOVERABLE OIL RESOURCES OF NEW ZEALAND AS OF DECEMBER 31, 1975
(In millions of barrels)

Rank/Field	Discovery Year	1975 Production	Cumulative Production	Reserves	Total Recovery (With Range of Estimates)
Others		1.4	7	118	125
New Zealand Total		1.4	7	118	125

NOTES: Although one giant gas field has been discovered offshore, New Zealand does not have any significant oil fields. Reserves of petroleum liquids are primarily condensate.

Table A.73

GIANT OIL FIELDS AND RECOVERABLE OIL RESOURCES OF PAKISTAN AS OF DECEMBER 31, 1975
(In millions of barrels)

Rank/Field	Discovery Year	1975 Production	Cumulative Production	Reserves	Total Recovery (With Range of Estimates)
Others		2.4	72	53	125
Pakistan Total		2.4	72	53	125

NOTES: The only significant petroleum finds in Pakistan prior to 1976 were gas fields. The recent *Dhodak* discovery added an estimated 200 million barrels to reserves.

Table A.74

GIANT OIL FIELDS AND RECOVERABLE OIL RESOURCES OF TAIWAN AS OF DECEMBER 31, 1975
(In millions of barrels)

Rank/Field	Discovery Year	1975 Production	Cumulative Production	Reserves	Total Recovery (With Range of Estimates)
Others		1.4	8	17	25
Taiwan Total		1.4	8	17	25

NOTE: Taiwan has no significant oil fields.